White Coat

White Coat

Becoming a Doctor at
Harvard Medical School

ELLEN
LERNER
ROTHMAN,
M.D.

Perennial

An Imprint of HarperCollins*Publishers*

First Perennial edition published 2000.

Designed by Sam Potts

The Library of Congress has catalogued the hardcover edition as follows:

Rothman, Ellen Lerner.
        White coat : becoming a doctor at Harvard Medical School / Ellen Lerner Rothman.
            p. cm
        ISBN 0-688-15313-5
        1. Rothman, Ellen Lerner. 2. Medical students—United States—Biography. 3. Harvard Medical School. I. Title.
R154.R755A3   1999   [B]   ·   98-38202
610'.71'17444—dc21   CIP

ISBN 0-688-17589-9 (pbk.)

09 ❖/RRD 20 19 18 17 16 15 14 13

To my patients,
for granting me the privilege
of sharing their lives and
for being my most powerful teachers

To Carlos,
my first and best reader,
without whom this book
would not have been possible

Acknowledgments

I first and foremost need to thank Carlos for giving me the strength and courage to write this book. I am grateful to my parents for believing I could do this and for their support throughout these four years. I am thankful for the interest and involvement Gregory and Debora Lerner have shown this project.

I want to thank my first readers, Lynn Gordon, Phyllis Mindell, Rosemary Mancini, Deborah Raub, and Jennifer Hillygus. Their early comments were instrumental to this book. I am grateful to Ms. Mary Ann Satter for inspiring my passion for writing and to Bob Whitaker for giving me my first opportunity. I want to thank my agent, Kip Kotzen, for noticing me, and my editor, Zachary Schisgal, and his assistant, Taije Silverman, for their excellent instincts and their understanding, supportive manner.

ACKNOWLEDGMENTS

I am honored to have shared this experience of becoming a doctor with my many talented, witty, and insightful classmates. They have taught me a great deal and have helped me process these past years.

I am in tremendous debt to my patients. I have been honored to share in their lives, even in a small way. My relationships with them have been powerful, thought-provoking, painful, and beautiful.

In an effort to protect the anonymity of the classmates, doctors, and patients I have worked with, all the names, and some details, in *White Coat* have been changed.

Contents

THE CLINICAL YEARS

White Coat

White Coat

You'll never *ever* guess what I did," Roy said over the phone. He had just returned from a clinic where he followed a physician as he saw his patients.

Roy was the first member of our class to perform a rectal exam. In fact, besides taking blood pressure, it was the first procedure any of us had performed. The gentleman Roy practiced on was subjected to three prostate exams on that particular visit—one from the physician and two from the medical students. But as uncomfortable as the experience must have been for the patient, it was equally awkward for Roy.

When I told my mother about Roy's experience, she was incredulous that the patient permitted such inexperienced hands to probe his prostate. "The patient actually *allowed* that?"

The only way to explain the patient's willingness was Roy's white coat. After several months of wearing mine, I was already accustomed to patient trust way out of proportion to my abilities. Another classmate questioned a patient about his diagnosis. Unfamiliar with the disease, he could only ask, "Um, do you think you could tell me more about what that is?"

The patient replied, "I was hoping *you* could."

My classmates and I received our white coats with "Harvard Medical School" embroidered on the breast in crimson cursive on the first day of orientation to medical school in our white coat ceremony. Our event in the Holmes Society was anything but ceremonious. Our class was divided randomly into four different societies, mainly for administrative purposes. Each of the four societies hosted its own ceremony, and we all met afterward for lunch, self-consciously checking one another out in the new and unfamiliar white lab coats. I stood near the end of a long, disorganized line in the Holmes Society office, waiting to receive my coat. By the time I reached the front, all the small coats had been given out, and I received one several sizes too large.

"You can trade with someone," the administrative assistant said.

A day later, wearing our coats still creased from the packaging, we attended our first patient clinic as formal members of the medical world.

The white coat ceremony, a new idea from the administration, was intended to herald our induction into the medical community on our first day of medical school. While not the long coat of a physician or resident, the white coat signaled our medical affiliation and differentiated us from the civilian visitors and volunteers.

This was not an affiliation I was ready to claim as a first-year medical student. Over the course of the year, after taking courses in anatomy, pharmacology, biochemistry, physiology, genetics, and embryology, I was more deeply impressed by how little I knew than by how much I had learned. Yet every Monday in our

Patient-Doctor course I found myself in my white coat interviewing still another patient.

Despite the uncertainty of my place in the medical world, my white coat ushered me into the foreign world of the patient-doctor dynamic. To my patients, the white coat denoted the authority and trust ascribed to physicians by the general public. Most patients were not attuned to the medical hierarchy designated by coat length. A white coat is a white coat is a white coat. Never mind that my coat loudly proclaimed "medical student." I felt as if I wore the scarlet letter, but no one knew what it stood for.

These weekly interviews as part of our Patient-Doctor course were about learning the important questions, the right mannerisms, and the appropriate responses to our patients. Our instructors taught us to take a careful, methodical history, which I more or less skillfully replicated every week with a different patient. Although the goal of these weekly patient interactions was to discover a person's experience of illness, these interviews were more about my learning process than about the patient's story. As I walked with my classmate back to the medical school from the hospital after a Patient-Doctor session, Andrea remarked, "I hate this. I'm so caught up in figuring out the next question that I can't really focus on the patient's story at all. Do you think this will ever change?"

When I interviewed patients, they saw my white coat. Many of my patients were well into their seventies, and at twenty-two I must have seemed a child to them. The white coat masked my youth. It masked my inexperience. It masked my nervousness. Yet in the medical world my white coat did not offer the solace of anonymity but forced me to take on power that I was not ready to accept.

As a white coat I could ask any question, and patients felt obligated to answer. They trusted me to hear their story without judgment, to understand their symptoms and their suffering, to listen with compassion. I collected information about their most personal problems and asked them about some of the most deeply private parts of their physical and psychological lives. In return they learned nothing about me.

Furthermore, these weekly interactions imposed power without responsibility. Every week I left the patients' rooms with a few pages of frantically scribbled notes, never to return. Their lives and our interaction were reduced to my chicken scratch. I had no relationship to the patients' care. My continuing obligation to the person was restricted to the requirements of confidentiality.

Before entering medical school, I would not have thought twice about allowing a medical student perform a rectal exam on me. The white coat would have fooled me too. While I fully appreciated the opportunity afforded me by these patients to learn how to interview and perform simple procedures, I looked forward to a time when I would be able to offer my patients concrete skills. I looked forward to growing into my white coat.

FIRST YEAR

Arriving at Harvard

I didn't remember the medical school very well from my interview visit the previous fall. I was so certain that I would not be coming here that I hadn't bothered to look at the campus very carefully. When I arrived for orientation in August of my first year, I felt as though I were seeing the medical school for the first time.

The looming white marble buildings of Harvard Medical School dominated Longwood Avenue. The five neoclassical buildings composing the medical school quadrangle were arranged around a long lawn with a few small trees laid out at regular intervals. Four identical buildings, two on each side, lined the long sides of the lawn. Each building had four panels of narrow green glass windows, and each pair of buildings was connected by a glassed-in atrium. The quadrangle buildings had been renovated in the late 1980s, and although one of the other buildings had been recently renamed for a donor, they were still known simply as buildings A, B, C, and D. Building E, better known as the MEC, the Medical Education Center, held most of the medical school classrooms. Building A took up the far side of the green. Set apart from the other buildings by a tall set of white marble steps, it was the most imposing of the five buildings with its row of three-story Ionic columns and "Harvard Medical School" engraved in the frieze above. It housed the president's office and the rest of the administrative offices.

Across the street from the quadrangle was a traffic circle. Avenue Louis Pasteur came off the traffic circle at right angles to Longwood Avenue, and two identical tan brick buildings sat on either side. They were oddly shaped pentagonal buildings and painfully ugly compared with the marble complex across the street. The second of the two tan buildings was the dormitory for the medical school, Vanderbilt Hall, fondly called Vandy, where I spent my first year.

The entry hall of Vandy was dim. The floors were lined with dark brown stone, and a dingy bronze chandelier in the circular ceiling cast a dull yellow glow. Two short hallways opened off the circular entry. The dark hallways were lined with row after row of gray-brown antique mailboxes. The back exit of the entryway led outside to the Vandy courtyard.

I left the courtyard and climbed a flight of stairs to my hallway on the second floor. I passed a communal kitchen on my left, and my room was just a few more doors down on the right. It was not large, but good-size for a dorm. It had hardwood floors, and the only window in the room overlooked the traffic circle. There was room for a bed on the left, and a small set of shelves next to the desk and a tiny closet completed the room's furnishings.

As I unpacked, I was still amazed to find myself at medical school at all, and at Harvard Medical School in particular. Medicine was a late discovery for me. I first thought of applying to medical school at the end of my sophomore year in college but didn't make up my mind until the middle of my junior year. I scrambled to finish all the required premedical classes in time to apply for the class entering in fall 1994.

When I started college, I planned to major in the classics and follow that by a career in law, preferably constitutional law. As a freshman I took an introductory biology course for my science requirement. Much to my surprise, I loved it. I also realized I had absolutely zero talent for Latin. As I became more interested in scientific issues through my biology courses, I started writing for the *Yale Scientific Magazine* to explore political and social issues in the sciences. I became particularly intrigued by medical ethics, and in my junior year I decided to apply to medical school to prepare for a career in medical ethics. I wasn't sure I wanted to practice clinical medicine, and I even considered a joint degree in law.

As excited as I was by medicine and medical ethics, I hated the premed mentality. The competition for admission to medical school was fierce—nearly forty-five thousand students applied for sixteen thousand positions each year. But the competition actually began much earlier in the so-called weed-out courses of organic chemistry, physics, chemistry, and introductory biology. Some students sabotaged classmates' experiments and hid required materials in the library to achieve better positions on the grading curve. I thought people were obsessed with their grades and devoted endless energy to tailoring and padding their résumés with activities they thought would enhance their applications.

My worst premed experience was hospital volunteer work. In the first semester of my sophomore year I decided to volunteer at the hospital to explore what practicing medicine was like. I ventured over to Yale–New Haven Hospital one early fall evening for the organizational meeting. I was surrounded in the small auditorium by nearly fifty premeds. After the introduction, several hands immediately popped into the air. The coordinator called on someone sitting near the front, who asked whether volunteering would help him get into medical school. The coordinator said that indeed medical schools expected to see at least one year of volunteer experience on applicants' résumés. My classmate sat back and nodded his head, relieved.

"Any more questions?" the coordinator asked. No hands were raised.

"Okay, the sign-up sheet is in the front, and we'll call you and let you know your schedule," she said.

I decided to volunteer in the respiratory therapy department. I had just finished a summer of research in a neonatal pulmonology lab and wanted to see the clinical side of my research. But what I actually did was help stock the various respiratory therapy closets around the hospital with ventilator supplies one afternoon a week. I was assigned to help Diane, a terse obese black woman with whom I was never able to establish a rapport. We started in the basement, where we filled a large cart with various types of plastic tubes and connectors. Then we worked our way through the hospital, filling empty bins. Every fifteen minutes she stopped for a twenty-minute coffee break. At each break she filled a styrofoam cup two-thirds full with stale coffee. Then she grabbed ten blue packets of Equal sweetener from the container next to the coffee machine, and taking five at a time, she carefully shook them down and emptied them into her coffee. Then she downed the entire cup. I lasted three or four sessions before giving up on volunteering altogether.

Applying to medical school was a misery. I wasn't confident I would get in, so I applied to nearly twenty schools. Each application had two parts to be completed separately, each school required several essay subjects and very few of the essay subjects overlapped. Traveling to the various schools was time-consuming and expensive, and interviewers found many ways to make students nervous. There was the standard lore about the interviewer who asked the student to open a window he knew had been nailed shut to see how the student handled the situation. The other traditional story was the interviewer who made the student wait alone in an office and then called his own phone to see if the student answered. If the student answered, the interviewer reprimanded her for answering another person's phone; if she didn't, he complained she let the call go by without taking a message. It was a lose-lose situation. Although I've heard these stories repeated often, I've never actually met anyone to whom this has happened. One of my Yale classmates, however, did tell the story of her overweight interviewer, who asked her point-blank, "Do you think I'm fat?"

I had my own nightmares. In one particularly bad interview I was called into an office by a man in late middle age. He was thin with straight silver hair, a few wrinkles around his brown eyes, a deep crease in the middle of his forehead, and thin flesh-tone lips.

"Oh, I see you're interested in ethics," he said to me at the beginning of the interview. I smiled a little awkwardly.

"Well, let me tell you about my dog. I've had my dog for a long time. She's ten years old, and my kids love her. But now both my kids are out of the house, and the dog has developed diabetes. We have to check her blood sugar and give her insulin. It's a big hassle, and she really hates the shots. And now that the kids are gone, my wife and I were hoping to travel. But we can't do it with this sick dog. Should we put the dog to sleep?"

"Well"—I waffled—"it depends on what kind of life you can give your dog and what you feel is appropriate quality of life for the dog. . . ." I was sure he was looking for an answer, but I had no idea what he wanted to hear and what would offend him.

"Naah, that's too easy. Let's make it harder. My wife's father became very ill recently. He was a healthy guy. He always said that if he got sick, he'd rather commit suicide than live incapacitated. So a few months ago he gets cancer, and it spreads to his bones. He's in terrible pain, and he asks me to give him a shot of morphine. Should I have overdosed him?" he asked.

"Well," I said, "it depends on what his state of mind was after he became ill. It depends on whether he was able to voice his own concerns and whether you thought he might be depressed. . . ."

"That's not an answer," he said.

"Well, you need to consider all the variables——"

"But that's not an answer. That's the problem with all you ethics folks. You have all these highfalutin ideas, but when it comes down to it, you have no answers," he said.

"Well, if it were me, no matter how much my father wanted me to help him end his life, it would be too hard for me to live with the fact that I was the one who depressed the syringe and gave him the overdose," I finally told him. What I really wanted to say was, "Kill the dog, kill your father-in-law, and stop bugging me with your stupid questions."

I remember clearly the day I received my acceptance letter to Harvard. One afternoon in early March I went to the post office and found three envelopes sitting in my box. I had a thick packet from Duke, a thin letter from Harvard, and a credit card solicitation. A thin envelope was never a good sign, and I had already prepared myself for rejection. I kept reading the letter looking for the phrase "due to an outstanding group of applicants, we regret that we are unable to offer . . ." As I puzzled over the thin letter on the way into my dorm, a classmate came up behind me and noticed the Harvard insignia on the letterhead. "Is that from Harvard?" he asked.

"Yes," I told him.

"In or out?" he asked.

"What?"

"Are you in or out?"

I looked over the letter one more time. "In, I think."

If Harvard didn't have anything more to say to me, they could have at least stuffed a few pieces of blank paper in the envelope to make it appear a little more promising.

The elation came only later. When I received the letter, I was unable to reach either of my parents, and I spent the afternoon in my dorm, trying to focus on my lab report due the following morning. In the evening I had several excited phone calls with my family as the realization that I had actually been accepted set in.

Sitting that first afternoon in my room in Vanderbilt Hall, within minutes of meeting my new Harvard classmates, I couldn't believe my luck. What were they thinking when they accepted me? Was it a mistake? I was terrified. My classmates would be brilliant. How would I measure up? What if medicine was the wrong choice for me after all? I tried to push aside my fears as I went back outside to the tennis court for our welcoming barbecue.

FEAT, the First-Year Education Adventure Trip, was a five-day orientation hiking trip. Roy, one of my best friends in medical school, was in my FEAT group. He was not exactly the outdoorsy type, and I'll never understand why he chose to participate. On the very first day of our hike the bus pulled up to the side of the road and dropped us off. We could still see the yellow school bus as we crossed a small stream approximately four feet wide and five inches deep. We all did fine, making our way across, until it was Roy's turn. He promptly slipped off a rock and stepped right in the stream, soaking his boot. On the way back across, he slipped again and stuck his other foot in the stream. He complained for the next four days that his boots never completely dried. When we finally reached the parking lot where the bus was waiting to pick us up at the end of the five-day trip, Roy, thrilled to be out of the woods, ran up and kissed the bus. When it came time to sign up to be FEAT leaders for the incoming class at the end of our first year, Roy said he would sooner die. No, he said, he would start his own orientation trip, FEAST—First-Year Education Adventure *Shopping* Trip to the outlet malls of Kittery, Maine.

Roy was very tall, and his thinness accentuated his height. He had dark brown hair and soft brown eyes with angular features and a deep cleft in the middle of his chin. By far his most notable feature was his loud piercing voice. I had to hold the phone a few inches away from my ear when he was on the line. His mother had sent him for hearing test after hearing test as a child, convinced that if he talked so loudly, he must not hear well. But one after another the hearing tests turned out normal. It was just his voice.

I met Carlos through FEAT as well. He wasn't in my group, but on our first night back in Boston, a group of us decided to go to Cambridge for dinner. A few people from other groups joined about half of my group for the outing. Carlos sat across from me at the table, and we talked some at dinner and a little more on the subway back to the medical area. When we got back to Vandy, we all split up and went back to our own rooms. About an hour later I bumped into Carlos in the hallway, but I didn't recognize him from dinner and introduced myself again. He managed to forgive me, but he never let me live down that social blunder.

Carlos moved to California from Argentina when he was eleven, and he still had a slight nasal accent to his English. He was on the shorter side, with light brown hair and olive brown eyes. He wore glasses with thin, oval tortoise frames. He had a gentle way about him and a quick wit. He had recently returned from England, where he earned a master's degree in health economics with the support of a Marshall scholarship. One of the most talented students in our class, he always knew the right answer but was never condescending. Carlos and I became close friends during first year, began dating at the beginning of second year, and became engaged at the end of our third year.

After the initial exhausting weeks of frenzied excitement and anxiety, my life quickly settled into an easy rhythm. I woke up each morning by seven forty-five and bumped into my classmates in the hallway outside my room and the Vandy entryway as we rushed across the street to the MEC for our eight-thirty lecture. The lecture hall had been recently renovated, and we sat in steep rows reaching up two stories. The more eager of our classmates arrived early to claim seats in the first and second rows, and by the end of the first month the rest of us had also staked out our preferred seating areas. Roy was usually a few minutes late and had to sit in the back of the lecture hall. After the lecture we went upstairs to the skills areas, where we had tutorials to discuss cases in small conference rooms or slide sessions in the bays of microscopes.

After the morning sessions were over, we lingered for a few moments in the atrium and society offices where we had met on the first day for our white coat ceremony. I usually stopped to chat with Carlos or Roy and then headed back to my dorm for lunch. On Mondays I went to one of the hospitals for my Patient-Doctor session in the afternoon, and I had classes on Tuesday and Thursday afternoons as well. I generally made it back to my dorm in time to squeeze in a run before dinner. The other afternoons I spent studying in my room.

My friends and I gathered for tea every night at eleven-thirty. It was a tradition I had started in college and brought to my friends in medical school. Most of the time we met in my room, but occasionally we rotated. On Thursday nights we gathered in Roy's room to watch the NBC series *ER*. We usually combined tea and TV on Thursday nights so we could still get a little work done afterward. Carlos, Roy, and I were the core regulars. We met for tea every night. Sometimes it was only the three of us, especially when we were close to exams, but usually a few others dropped by, and we were generally a crowd of four to eight people. Tea was by far my favorite part of first year. We talked about everything from who had fallen asleep in lecture that day, to the topics we studied in class, to our personal lives.

Weekends were generally quiet. Many classmates, including Carlos, traveled to see out-of-town significant others. I usually studied intensively on Sundays. If we had an upcoming exam, I studied on Saturdays as well. But more often I spent Saturdays exploring Boston with my classmates. We went out to dinner and to the theater or walked along the shops of upscale Newbury Street. I visited the Museum of Fine Arts, the aquarium, the historic sites along the Freedom Trail.

*   *   *

As I had anticipated, our learning experience was intense. According to the New Pathway system of education, a small portion of our time was spent in lecture halls, while the bulk of our learning came from problem-based small-group discussions where we taught each other. This new system of education worked extremely well for me. I, like many of my classmates, was a motivated learner and liked to create my own questions and discover my own solutions. The new system transformed the traditional rote memorization of medicine into an intellectual process. In the tutorials, with the help of preceptors, my classmates and I learned basic medical themes from specific patient cases; we also had small lab sessions where we learned pathology and histology.

My classmates and I followed a classroom-based curriculum for the first two years of medical school. Despite the presence of many faculty resources, the learning process was designed to be independent and self-directed. Rather than take on a prominent role in our education, the faculty supervised us and ensured that we assimilated the essential information. We switched topics rapidly, so very few lecturers or tutorial leaders were with us for more than several weeks at a time.

The first year of medical school was devoted to normal human physiology; in the second year we learned about diseases. Many of my friends at other medical schools envied Harvard's pass-fail grading system during the preclinical years, but the pressure was still intense. We received numerical grades on each exam, but our administrators recorded only whether we passed. We were not formally ranked in an effort to spare us the competition. But we were shown the curves, and although we never knew exactly who had received the highest grade, we could easily place ourselves among our classmates.

We entered the hospitals for the more rigorous portion of our training in the third and fourth years. During these second two years we did rotations, in which we worked in the hospital with doctors and residents in the different fields of medicine, including surgery, pediatrics, obstetrics and gynecology, psychiatry, radiology, and neurology. During these years, unlike the first two, we received grades for our clinical rotations. These grades ranged from high honors through honors to satisfactory and finally to unsatisfactory. They were determined by the residents and senior physicians my classmates and I worked with on the hospital wards. Some rotations had tests, but many didn't, and these grades were routinely criticized for being subjective. Often they were completed months after we had left particular wards. Far more than the grades we earned during the first two years, these marks determined our standings as we applied for residency positions.

The greatest innovation of the New Pathway was our Patient-Doctor course. This was a three-year curriculum intended to inculcate humanistic medicine and to introduce us to patients at the outset of our medical education. The first year was devoted to instructing us on how to take a thorough medical history, but we spent the bulk of the time as our first patient experience, learning how to relate and respond to our patients. We met one afternoon per week for two hours, and we spent almost every session interviewing a patient, either as a group or individually, and then we discussed our experiences. We had several videotaped sessions where we could watch ourselves and evaluate our interactions. In the second year we learned physical exam skills. We continued to meet weekly for the first portion of the year and examined patients during each session. As the year progressed, we spent more and more time in the hospital, until Patient-Doctor became two full days per week in the last few months of the year. During the third year we went back to a once-weekly schedule as we dealt with clinical and personal issues we encountered on the hospital wards. With a few exceptions we stayed throughout the three years with the same small groups we were assigned to in our first-year patient-doctor session.

The first year of Patient-Doctor introduced me to the primary challenges I faced as I progressed through my four years of medical school. While these challenges grew in complexity as my clinical sophistication developed, the seedling of each already thrived in my first experiences. When I arrived at medical school, I thought the academic challenges and the grueling hours would make it difficult. Yet as I quickly adjusted to the pace, I soon realized these were actually the easiest hurdles to surmount. For me, the greatest challenge was to accept the responsibility of being a doctor and to grow comfortable with the intimacy of the medical relationship. I had always considered the patient-doctor relationship a true partnership, but I quickly began to recognize the power discrepancies inherent in the situation. I had thought confronting death would be hard, but I soon learned dying was harder. Before medical school I always had been able to master the tasks set before me, but now the goals seemed far vaster than I could ever achieve. I had always assumed that I would marry and have a family, but now I recognized the challenge an intense medical career would pose to this ideal.

Before coming to medical school, I thought I knew what I meant when I said, "I want to be a doctor." But as I took my first steps in medicine, I realized I had no idea.

Anatomy Lab

We don't know who they are. We will never know why they chose to be there. To them, we are Medicine. To us, they are the Human Body.

For me, anatomy lab presented my first real experience with death. On the first Monday of medical school I faced a roomful of dead people who had chosen to donate their bodies to medicine. The sweet scent of formaldehyde seeped out from behind the gray metal door, and I hesitated for a moment before I boldly pushed it open with my clammy hands. I entered into the middle of three bays of anatomy labs. Steel tables covered with indigo cloths lined the walls. The drape of the cloths vaguely suggested the human shapes beneath. The far wall was all windows, and the warm morning sun streamed in, throwing shapes and shadows onto the denim forms and the speckled cream linoleum floor. The wide doors between the bays were opened, and as I looked to my left and my right, I could see rows of denim gurneys extending in the other identical rooms. I turned left and walked into the far bay to find my cadaver. It was second to the last on the left, indistinguishable from all the other veiled forms.

On that first morning my four classmates and I pulled back the denim sheath to reveal the body beneath. We discovered a woman. Her small, wrinkled breasts with pink-brown nipples fell slightly to the sides. Her head and hands were wrapped in muslin secured with a large metal safety-pin. Too timid to invade her privacy further, we exposed only her torso that morning, carefully replacing the blue denim to hide her again. Over the course of the two-month lab class I became better acquainted with her body and developed a more detailed understanding of her anatomy. I traced her splenic artery to the celiac trunk, her median nerve through the brachial plexus, and the entire length of her gastrointestinal tract.

Yet my introduction to death felt backwards. I was thrust into the physicality of death without experiencing the human struggle of the dying process. I knew my woman's body more intimately than that of any other living person, but I knew nothing of her spirit, her anima.

The preparation for the lab had stripped away even the most basic clues to her personhood. The fixative obliterated her years and made her ageless; the dissection process chipped away at her body to make her formless. Her face was sealed away in its muslin shroud.

At the memorial service for our cadavers, Alyssa voiced her struggle to make the transition from the physical body to the uniqueness of the individual. "Every time we looked at a new structure," she said, "I thought, This is the one that's going to be different. This is the one that's going to make him special. Yet every time it was the same. All our bodies looked the same."

Almost all of us dreamed about our cadavers at some point during the anatomy course. Dean Ed Hundert forewarned us; it was common. I was prepared for nightly visions of scalpels dissecting and gloved hands retracting. But my recurring dream was about holding my cadaver's hand. In the dream the five of us stood in our customary positions around the gurney, obviously at work on the body. I held her hand as one might to reassure a friend or relative undergoing a painful and frightening medical procedure. In my dream I forged the human connection and bond that could not exist in the lab.

Despite the frustration at the lack of personal connection with our cadavers, I didn't want to know any more than I did. It would be too painful to dissect structure after structure with the burden of personhood and the relationships and responsibilities it entailed.

But if my first experience with death was with the physical, I soon confronted the emotional, human aspect. In my Patient-Doctor course I was assigned to interview Sara on Monday afternoon the week before our anatomy course ended.

Sara was seventy-four, dying, and alone. When I came into her room, she rested quietly in her bed, arms folded across her stomach, looking toward the blank TV screen. The aggressively neutral curtain with its nondescript, faded-away pattern was drawn to separate her from the patient in the bed next to her. Sara was close to death from congestive heart failure after having outlived a one-year prognosis by eleven years. She was angry that she was dying, although she wanted the inner peace acceptance would bring. As she became sicker and less mobile, she was increasingly isolated from her friends and alienated from her family. Frustrated by doctors who no longer made time for her, she attributed their brusqueness to her rapidly deteriorating medical condition.

"I understand that it's difficult for them to get close to someone who is dying. But when they don't say anything to me, it means either they have nothing supportive to say as physicians, or they don't care as people," she said.

There have been others. Amanda was sixteen and in the terminal phase of AIDS. Several classmates and I interviewed her together as part of our Patient-Doctor I course. Amanda lay passively on her bed watching TV, her brown-blond hair drawn into an impromptu ponytail. Her lips were covered with cracked maroon crust, and the catheter tip of her central venous access protruded from the neck of her johnny top. She had watched her mother and her younger brother die of AIDS the previous year. No longer able to eat, she had only the occasional craving for pretzels. She had given up on school more than a year before. Now she mostly watched TV all day. We asked what her favorite show was. "I don't know." Her favorite actor: "I don't know." Her favorite thing to do: "I don't know." She had already begun the process of relinquishing her personhood.

Dan and Steve also had AIDS. I visited them at their home through the elective course Living with Life-Threatening Illness, which exposed students to the issues of terminal illness, pain management, and hospice care. Steve was in bed when I arrived, but Dan carefully helped him to the living room and tenderly arranged him in a rocking chair. While Dan still seemed healthy, Steve's gaunt features, watery blue eyes, and deep shadows gave him the unmistakable look of chronic, terminal illness.

When we met, Steve and Dan had already stopped setting long-term goals for themselves; they were in the midst of planning an auction since they believed their family members would not care for their antiques properly after they died. Yet even as they planned for their own deaths, they still set ambitious objectives for their cat: Inky was on a new diet.

I frequently thought about all of them. But as Christmas of my first year approached, Sara dominated my thoughts. Her physicians didn't expect her to live past December. When I planned my trip home for winter break, I wondered if she was still alive. Was she able to reconcile with her family? Was she able to forge a more meaningful relationship with her friends?

I found myself drawn to these dying people. I was taken with their struggle to be human and to be better humans in the face of death. As I became more deeply involved in the lives of these patients and their care issues, I developed a passion for the clinical encounter and the patient-doctor experience. Not sure I wanted to be a practicing physician when I arrived at medical school, I now realized I could not tolerate a career in ethics, abstractly contemplating patients' lives. I wanted to be enmeshed.

At times it felt as if death were everywhere. In anatomy lab we finally uncovered the facial shroud and opened the skull to dissect the brain, and that was okay. I talked to a patient who had nearly died the previous evening and would certainly die within the next months, and that was okay. I came home, and my goldfish had died, and that wasn't okay. I sobbed for half an hour.

ER

*E*R. Around here it wasn't an option. It was an obligation. Every Thursday at ten, we gathered in the designated rooms: Roy's room, Marisa's room, the Culver Lounge. It wasn't a question of whether to watch, only where.

I always watched in Roy's room. I tried to get there early to get a spot on the bed; latecomers sat on the floor.

My classmates and I started watching the series out of loyalty to Harvard. With HMS graduate Michael Crichton and third-year medical student Neal Baer writing a story line based on their experiences at the Massachusetts General Hospital, how could we miss it? The programs were rumored to follow our curriculum because cases on the show often corresponded to the topics we studied in class. But *ER* immediately transcended our sense of duty. We were hooked.

For me, *ER* was not just another popular TV show. It was the experience of watching physicians, residents, and medical students deal with detailed medical information against a backdrop of complicated personal situations and ethical issues. It was like watching seven or eight live-action tutorial cases in an hour. But more than the medicine, it was the excitement of watching my appreciation of the show broaden as my understanding of the clinical issues and the dynamics of the patient-doctor relationship deepened.

I remember the first time I actually understood some of the medicine on the show. It was during anatomy, and we were in the middle of a cardiac case. I don't even think I was totally clear on how blood circulated through the heart. But when that patient on *ER* came in with the emergency medical technician from the ambulance yelling, "V fib—we're gonna shock her," our room echoed with "V fib—that's *ventricular fibrillation!*" We all knew that sending an electrical current through this woman's chest was intended to send her cardiac cells into the refractory period and electrically reset them so her heart could beat synchronously again. It was a moment of arrival symbolizing our induction into the medical community.

As we moved through our curriculum from anatomy to physiology and pharmacology, the experience of watching *ER* changed. In physiology I was captivated by the man with congestive heart failure. During pharmacology I was interested in the drugs more than anything else. When I learned that "IV push" meant bolus injection, streptokinase was a declotting agent, and lidocaine was an antiarrhythmic, I had another epiphany in our acculturation into the medical world: I could finally decode the blur of medications constituting the bulk of the rapid-fire jargon on the program. As I added more classes of drugs to my repertoire each week, my decoding abilities increased exponentially.

ER transcended the dorm and pervaded the classroom as well. One of my classmates gave her tutorial group a detailed description of an EKG (electrocardiogram) machine, its readout of the heartbeat pattern, and the type of information it measured. Impressed, another student asked where she had learned about the machine. "Oh," she said, "I saw it on *ER*."

It wasn't the medicine alone that drew us back to the same rooms to watch week after week. We watched the physicians, residents, and medical students finesse or flub difficult situations with patients and their families, and we reflected on how we would have responded. When the arrogant surgical resident Benton missed an appendicitis diagnosis because he failed to listen to the patient, Roy announced to general agreement, "He certainly didn't have Patient-Doctor One."

Sometimes it's not how we would respond in similar situations but how we have responded. Watching third-year medical student Carter's first patient interview was as momentous as recognizing ventricular fibrillation. When that episode aired, we had also just done our first patient interviews. In the *ER* episode Carter tries to interview a senile elderly woman who only sings in response to his questions. He can't even get her to say her name. When the senior physician returns and immediately elicits the pertinent information, Carter is humiliated. At that moment there was a palpable silence in the room. No one said anything, but we all thought, That was me. The intimidation of talking to the patient, the pressure to get the "right" information, the frustration at our own lack of ability: That was me.

I quickly became an *ER* addict, watching religiously every Thursday night. Surprisingly, I found it very difficult to watch during vacations. While I found it enjoyable and relaxing on school nights, it was jarring and stress-provoking when I was on a break from medicine. The classes ahead of us and behind us did not have the same relationship with *ER* that we did. Perhaps *ER* paralleled our experience more; we started out together. We felt a strong bond with Carter that our more jaded upperclassmen might not have shared. But since it began during our first months of first year, *ER* easily became an accepted part of our school routine.

Through the *ER* physicians, residents, and medical students, my classmates and I explored who we wanted to be and what we were afraid we might become. We developed a paradigm for how we wanted to respond to our patients and explored how we would feel if we were unable to uphold it.

Taking a Sexual History

The situation was tense. It was the Patient-Doctor I final videotaped interview required to pass the course. Seth interviewed his patient-actor in a small realistic medical examining room set up in a far corner of the MEC for testing purposes. He knew his preceptor was observing and videotaping the entire conversation through the one-way mirror. He also knew we would review our tapes in our small-group discussion sessions the following week. And he had less than five minutes left in which to obtain the patient's sexual history.

"So, let's talk about sex."

There were few ways to do it gracefully. It was universally one of the most difficult topics to broach with patients and one of the most challenging to learn as a student.

We live in a society obsessed with sex. Advertisers tempt us with the promise of sexual fulfillment, and movies equate great

sex with true love. Yet at the same time, talking about the real-
ities of sex and sexual function is one of the great taboos. Not
only must we confront societal inhibitions, but we had our own
beliefs as well. When taking a sexual history, we negotiated the
cultural norms and personal values of our patients at the same
time. On a topic that may have been uncomfortable for both us
and our patients, we somehow had to convey an atmosphere of
openness, honesty, and sensitivity.

Because of the highly personal nature of the information, the
sexual history exacerbated the power dynamic of the patient-
doctor relationship. As a student I was not ultimately responsible
either for the care or for the information. My interviews were
about the questions, not the answers. While discussing chest pain
seemed benign, the topic of impotence or a patient's history of
sexually transmitted disease seemed far too personal. As a result,
it became more difficult to justify the questions I asked. Was this
just sanctioned voyeurism?

Yet as daunting as this topic seemed to me, some of my classmates were perfectly at ease with the topic. When we came to
the two-week segment on sexuality in Patient-Doctor, Andrea lit
up. She exclaimed, "I have been waiting for this *all* year! You
know me. This is my *favorite*!"

Andrea was an intensely animated speaker. She used big, emphatic hand gestures, and her speech carried a sense of urgency.
I always got caught up in her excitement and found myself drawn
to her, craving her attention. After reading a book on deviant
sexuality during college, she had decided to devote her life to
studying human sexuality and had spent several years before medical school working with women who sold sex in the legalized
brothels in Nevada.

Sometimes patients appreciated our directness. During our biochemistry course, we had a tutorial case about a patient with a
rare genetic disease, osteogenesis imperfecta. In this disease, pro-
duction of collagen, a protein that forms the substance of bone
and connective tissues, is defective. As a result, patients are born

with fragile bones. The course administrators invited a patient with osteogenesis imperfecta to speak to our class.

Sandy sat in her wheelchair in the front of the amphitheater. She was very severely affected. Now forty years old, she was barely the size of a three-year-old. Her body was mainly head and torso, and she had tiny legs that did not reach beyond the edge of the seat of her wheelchair. As we listened, Sandy spoke eloquently of the difficulties she had encountered with her disease. When she finished, she opened the discussion up for questions. Andrea's hand immediately shot up.

"What is sexual activity like for you?"

I cringed, and I felt my classmates sitting around me shift in their seats as well. How could Andrea ask a question like *that* in a situation like this? Where was her tact? If a discussion of sexuality was barely okay for a private patient interview, it was grossly inappropriate in front of two hundred strangers.

But Sandy was thrilled. "Give that girl an A!" she said. "You know, I am so glad you asked me that. People always look at me and assume that I can't possibly be sexually active. But I have a wonderful partner right now. He's gentle with me, and it's very fulfilling for both of us. He worries that I'll break a bone, but we're careful. Once he dropped me in the bathtub, and he was mortified. The look on his face . . . ! He was sure I had broken something. But I was okay. We still laugh about that."

With the advent of the movement to remove gender and sexual stereotypes from language, which the popular press has termed political correctness (PC), taking a sexual history became even more difficult. Although it had always been important to know whether a patient was gay or straight, married or single, active or abstinent, we now danced around these topics while eliciting the information for fear of unconsciously imposing our values on our patients, thereby alienating them. In some ways this reluctance to identify the issues honestly reinforced the taboo of talking about sex.

Often these questions felt as awkward to our patients as they did to us. One afternoon Roy attempted to learn his patient's sexual orientation by using the PC method. The patient looked at him blankly for a minute. "Do you mean, am I gay? No, I'm straight. Why didn't you just ask me?"

Although our teachers encouraged us to ask these questions of all our patients, they often seemed inappropriate. "I just can't ask a seventy-four-year-old woman if she is sexually attracted to men, women, or both," said Renu, voicing a reluctance many of us shared. When she actually worked up the courage to ask this question of a married woman, the patient was offended by the possible implication. "Why did you ask that question? Of course I'm not sexually attracted to women. I'm married with two children!"

Renu, also in my Patient-Doctor group, was beautiful. She had long, thick brown-black hair, and the black of her pupil was barely distinguishable from the rich brown of her iris. She had golden brown skin and a thin mouth that smiled infrequently. Renu was conflicted about her experience with patients. She often felt uninterested in their stories and wondered what career she would find in medicine. She occasionally questioned her decision to enter medical school. Taking a sexual history exacerbated these concerns.

Despite the difficulties of adjusting to discussing sex and sexuality with my patients, as the year progressed, these conversations began to feel more routine and slightly less invasive.

Robert was a thirty-year-old man in the hospital for an exacerbation of his asthma. The flare had almost resolved, and he was being discharged the next morning. A large man, his eyes, hair, and skin were the same brown-black, broken only by the sharp contrast of the white of his eyes and teeth. He wore a navy blue hooded sweatshirt over the standard pale blue hospital pajama pants and green Styrofoam hospital slippers. I had finished discussing his asthma and had already moved on to his other medical problems when I remembered that I had forgotten to ask him what triggered his asthma.

"Oh, when I smoke cigarettes or crack," Robert said nonchalantly. I tried to hide my surprise. He was so casual about it.

"Do you use any other recreational drugs?" I asked him.

"Well, I do a couple of bags of heroin every day. But that's it," Robert said.

In the course of the conversation Robert told me that he hoped to go to rehab the next day. His girlfriend, who had been an IV drug user in the past, had been clean for the last year. As we discussed his girlfriend, I questioned him about safe sex. "Do you generally use a condom?"

"Naah. I don't like them."

"Well, you know that because you use IV drugs, you are in a higher-risk category for HIV and hepatitis, which can be sexually transmitted. Condoms prevent transmission."

"Yeah. I had an HIV test once. It was negative."

"Well, even so, both you and your girlfriend are in high-risk categories. I know condoms aren't the most pleasant things, but they are very important for you to protect both yourself and your girlfriend. I can't tell you what to do, but I want to make sure that you understand the importance of a condom for you."

"Yeah, I know I should. But I don't think that's really going to happen."

"Well, just as long as you understand."

Later that week I described the interview to my mother. While I had moved on to other topics, my mom was still stuck on the condom part. "Just a minute. You were telling who to wear what? It's going to take a lot of time for me to get used to this," she said. At least in that particular interaction with Robert, I was already comfortable enough to discuss sex easily.

"Penis," "erection," "vagina," and "orgasm" still didn't roll smoothly off my tongue. I still heard my voice catch as I stumbled over the word "function" every time I asked, "Do you have any questions or concerns about your sexual function?" But I kept on asking in the hope that someday these conversations would become second nature.

Great Expectations

Mount Success was our nemesis. We were tired, hungry, and sweaty after hiking in the Adirondack Mountains all day. Nine of my Harvard Med School colleagues and I attempted to summit the thirty-three-hundred-foot Mount Success before nightfall. We failed.

The irony of the situation was not lost on us.

When I arrived in August, I eagerly anticipated a pass-fail, relaxed academic atmosphere, dreaded the fabled workload and stress of medical school, and felt intimidated by the brilliance of my classmates.

My concerns dissipated as I adjusted to the course load and became friends with my peers. The reality shock came from the high stress that low-pressure academics elicited.

Rumors circulated about my first-year class in particular, especially just before exams. "I heard that our class is the most stressed-out class that has come through in *at least* the last five years." This quote was attributed to everyone from someone's friend's tutor to Holmes Society master Dan Goodenough himself.

After preparing for a pass-fail exam by a marathon week of studying, I remarked to Roy how overzealous our continuous vigil in the library seemed for a pass-fail exam. While we received numerical grades, all the administration recorded was whether we passed. I had never studied this hard for an exam that *was* graded in college.

"I know I'm going to pass. That's not the issue," he said. "I just want to pass with dignity."

We all joked about the "Super P" for the student who passes extraordinarily well. Yet we all strove for it. No matter how inconsequential each exam and how forgiving the administration, we expected ourselves to surmount and to succeed. For many of us, it was not enough just to pass.

At an open forum a classmate voiced her frustration about failing a course final. "Why is it that no one talks about failing a test? I saw the curve. There are ten more people down there with me. This is not something we should be ashamed of." In medical school, where we transgressed some of the most ingrained social mores, failure was the ultimate taboo.

We constantly scrutinized ourselves to maintain a safe distance from even the risk of failure. In a curriculum with no requirements, conversation was punctuated with "Are you going to learn that?," "What have you read?," and "Do you think that's important?"

Not only did we doubt ourselves, but we questioned the New Pathway and the learning process itself. Classmates worried that we did not learn enough facts and basic science. Most of us have heard stories of disparaging remarks about our scientific grounding by clinicians during the hospital rotations. We joked about reading *Harrison's,* the two-thousand-page tome of medicine, cover to cover. Rumor had it that one of our classmates actually did.

This collective doubt was contagious. Kate was my lab partner, and she quickly became one of my closest friends. She was a soft, warm person with long, very curly brown hair and a ready smile. She was predisposed to hugging, and her taste in clothes tended toward tie-dye. She was well known in our class for her knitting. She carried a bagful of colorful yarn, and anytime there was a lecture, she brought out her needles and knitted as she listened.

Kate came away frustrated and stressed from her tutorial after their discussion about the New Pathway. Her group had voiced the usual nagging doubts about the quantity of our education. Kate, who was generally pleased with our education, said, "Well, maybe I'm not learning enough. Maybe I'm not going to be a good doctor. But I guess I'm just going to have to trust that Harvard knows what they're doing," she said.

Medicine both rewarded and required this constant self-evaluation and desire to succeed. The glamour of medicine was the ability to save. According to this mystique, only partially grounded in reality, if I learned medicine's secrets, I would heal people suffering from "self-induced" diseases like alcoholic cirrhosis and smokers' lung cancer. I would rescue others from their genetically predetermined fates. I would understand the human body, the enigma of life. In turn patients would look to me to provide the daring and even dramatic escape from destiny.

At the same time, I must be above human fallacy. I must surmount fatigue and eradicate oversight and error. My mistakes and failures could have catastrophic consequences.

This idealized vision of medicine and health care was both exaggerated and self-aggrandizing. Nonetheless the pressure to learn everything was overwhelming.

In a pathology lab our instructor displayed someone's internal organs riddled with the mealy yellow tumors of lymphoma. "You must be extremely careful when palpating potential tumors in your patients," he said. "You may dislodge loosely associated cells which can then spread and produce new metastases." It was overwhelming to think that even my innocent touch could have deadly consequences.

We applied the same exacting standards to our personal lives as we did to our academic pursuits. We strove to stay one step ahead of failure, but sometimes we felt ourselves fall into the abyss. The struggle was fiercely individual for each of us, but to be sure, it raged in each one of us and left its fibrous, mottled scar as a reminder.

Late in March of first year one of our classmates attempted suicide. He and his girlfriend, also a classmate, had been together that morning in her dorm room. When she left to take a shower, he disappeared. He rented a car and drove to a bridge. He pulled over to the side of the road, left the car, and jumped to the bay many feet below. Only a miracle saved him.

"Right after he jumped, he realized that he had made the biggest mistake of his life," his girlfriend later told us. "He turned himself in the air and reoriented himself so that he hit the water rather than the concrete pier below." He survived the fall with only a fractured rib.

I couldn't believe what he had done. I had shared the biochemistry and pharmacology tutorials with both him and his girlfriend. We had met in our small group of eight students three times a week for two hours each session. I hadn't noticed him struggling, and I hadn't even suspected his depression. He was always well prepared for tutorials, knowledgeable about the topics, eager to participate. While he was never someone I spent a tremendous amount of time with outside class, we were friendly. I never saw even a hint that something was wrong.

Soon after the suicide attempt Alyssa received anonymous hate mail from classmates. I bumped into a group of my classmates in the atrium of the MEC. Alyssa sat in the center of the group, her eyes flaming and glassy with barely restrained tears. She held a sheet of paper and passed it around for the rest of us to see. The unsigned typed letter blamed Alyssa's competitive nature for damaging our class spirit and contributing to the atmosphere provoking our classmate's suicide attempt. The authors said they didn't appreciate her condescending attitude in tutorials, and quite a few people refused to be in a tutorial with her again. If she didn't modify her behavior, the letter warned, then the authors hoped she would leave our class.

Alyssa played a difficult role in our class. She was an older student, and I always thought she was beautiful. A bold speaker, she nevertheless had delicate gestures and soft movements. Alyssa worked very hard at learning medicine; during the second year she carried a black notebook in which she had painstakingly outlined *Robbins*, the thousand-page basic pathology reference book. The same attributes that made her such an exceptional student, however, could make her a difficult tutorial mate. She could be aggressive in anatomy lab to do most of the dissecting, to identify the structures herself, to get her questions answered first. In this respect she was certainly not alone in our class, and perhaps people judged her more harshly because she was a woman.

Socially I found Alyssa very pleasant, although I didn't spend much time with her outside class. She was exceptionally bright and always had interesting insights into patient experiences that she willingly shared during small-group sessions. Nonetheless she had few friends in the class, and she mostly kept to herself.

We never found out who wrote the note. I couldn't imagine who could have done something so cruel and cowardly. As inappropriately as Alyssa may have acted at times, nothing she did deserved this treatment. While the episode blew over in time, Alyssa became even more closed off from the rest of the class. She often ate her lunch alone, reading the paper in the society offices.

In the weeks following the suicide attempt, other members of our class came forward and shared their own trials. Dan Goodenough, master of the Holmes Society, sponsored an open forum for our class after the suicide attempt to help us process our feelings. A reformed hippie, he wore patterned Polartec and Birkenstocks while his colleagues wore suits and bow ties. If there was emoting to be done, Dan was on top of it.

I was surprised to find that many of my classmates also struggled with depression. People I would never have suspected came forward and described their sadness, their therapy, their relief through medication. One classmate stood and told us he was gay. He had been working hard to accept himself as a gay man, and he wanted us to embrace him as his complete self rather than continue to deny the issue. In the following week another classmate also disclosed his gay identity and then came out to his family as well. Still other classmates shared their battles with eating disorders. Most of us felt very stressed from the academic pressures and doubted that we could ever become competent, confident physicians.

I thought we knew one another well. Yet I was stunned to discover how little we actually understood one another. We had commiserated about the work and the stress, but I was horrified to learn how little support we had actually provided one another. At the meeting one person said, "I look around me, and I think that everyone else leads such a perfect life. Everyone else deals perfectly with the stress. It never gets to them. I thought I was the only one." Even among the people with whom we shared the experience of becoming a physician, we feared to acknowledge our private struggles, our perceived weaknesses.

Perfection. Medicine expected it. We demanded it. We failed in our first attempt at Mount Success. But we'll spend the rest of our lives trying.

Hospice

I turned the corner out of the elevator. The afternoon sun reached through the window at the end of the dim hallway and cast an exaggerated reflection of the window on the linoleum floor. Brown doors lined the dingy white corridor: 708, 709, 710. Room 712 sported an emerald green clover proclaiming to passersby, "I'm Irish." Room 716 was a brown door, just like all the rest.

I knocked softly on the door, and it opened just wide enough to let the aunt and the laundry out. I caught a glimpse of an obese woman sitting on the edge of a bed. She clasped a hand attached to an emaciated lower arm. The arm disappeared behind an apartment wall that hid its owner from peering eyes. The door quickly shut again.

"It won't be long now. His mother is with him now, and the nurse just left. Hours, at the most," his aunt explained. We took the laundry and headed back down the hallway to the elevator and the laundry room below. "If you could just put this in the dryer. You don't need to come back up." She left me with some quarters and disappeared into the elevator and back to the apartment in the hallway upstairs.

I found out later that Thomas died of AIDS three hours after I had deposited the clean and folded laundry on the linoleum floor next to the white box in the hallway. He was thirty-three. I was told that he had a very peaceful death.

His brown door looked so innocent, lined up with all the other brown doors in that corridor. It yielded no hint of the dying it concealed. I wondered what the other brown doors hid in the rooms behind them. I wondered if pain and death lurked behind their brown confines as well.

I spent my summer working with a hospice, an organization that delivers health care to terminally ill patients and their families. Hospice nurses, physicians, social workers, and volunteers traveled to patients' homes to provide medical care and counseling. They helped patients spend as much time as possible at home and helped decide when hospitalization was warranted.

In addition to spending time with the nurses and physicians making home visits, I volunteered to sit with patients and their families and support them through the dying and grieving processes. There was no specific goal for these meetings. Occasionally I drove a patient to a doctor's appointment or picked up a piece of medical equipment to deliver to his home, or I talked with patients and helped with simple household tasks. Despite my various small roles at the hospice, as a medical student I was in an awkward position, neither caregiver nor friend. I struggled to find a balance in my relationships.

Defining the boundaries in our patient relationships was difficult for all the members of the hospice team, not just me. I was already familiar with the struggle the medical caregivers faced. Each of us trod a poorly defined line between caring too much and dispassionate aloofness, but in hospice care, this conflict was even more acute. All our patients were dying, and our goal was to provide them a humane and peaceful medical experience at the end of their lives. We spent many team meetings discussing how to help a patient attend a wedding, achieve a reconciliation with a loved one, or come to peace with her situation. We discussed our patients' greatest fears and their hopes for their deaths. We addressed their pain management needs and plotted which procedures we could possibly do in their homes. We became deeply involved with each patient, and our patients, with little energy left to expend on frivolous social connections, formed deep bonds with us. It was hard to create distance in our caregiver-patient relationships.

For me, that summer was the first time a patient of mine died. While I had met through Patient-Doctor several people who were dying and had developed a meaningful and long-term relationship with Dan and Steve during my course Living with Life-Threatening Illness, none of them had actually died. Even after I had experienced the physicality of death in anatomy lab, death was still an abstract concept for me at the end of my first year of medical school.

David was still a young man, in his early fifties. He was small, balding, with wiry red wisps brushed across his broad forehead. He was devoted to his Russian Orthodox faith, and his life was focused on his family and his religious community. His wife, Joanne, was a soft-spoken woman, overweight, with auburn hair that hung to her waist. She padded through the house in worn slippers, interrupting us to offer David ice for his feet or morphine for his pain. They had six children. Melanie, the youngest, was only nine.

David was newly dying. Three weeks before he was admitted to the hospice, he became extremely fatigued while driving on a family vacation to Florida. When he returned to Boston, he visited his physician, who discovered a malignant melanoma, a skin tumor, that had metastasized to his liver. The dying process progressed quickly for David, and he was very ill by the time I met him. On my first visit we sat around his cluttered dining room table, he with his legs propped on the chair next to him in an effort to reduce the severe swelling in his feet. We pushed away the books and papers to make room for my pad of paper. Joanne brought me tea and David a glass of water. As we sat, he told me the story of his life that I was to record into a book as his last gift to his family.

On my second visit, only a week later, David lay in a darkened room, no longer able to sit up. I pulled up a rocking chair next to his bed. Whereas the stories had carried a logical thread the week before, now they were disconnected and difficult to understand. Joanne often came in to refresh the cold washcloth on his forehead, and she interpreted his words and fragments of sentences. Before I left, he gave me a card with the Russian saints. We planned to make a videotape on our next visit.

But David died only three days later. I wasn't surprised that he died, and I felt strangely calm when I heard the news a few days after his death. He had been admitted to the hospice because he would die soon, and that understanding made it easier for me to let him go. As a student I also hadn't felt personal responsibility for his medical care. Yet I had expected a stronger reaction. I wondered if I had set my boundary too close to the extreme of the medical automaton. I sent Joanne a condolence card along with the transcripts of my two conversations with David.

When I told my friends and family about my summer experience with the hospice, they often asked how I could handle it. "Isn't it too depressing?" my mother asked when I planned the summer project.

If the hospice were exclusively about dying, it would be too painful. And if it were about becoming inured to the death process, that would be painful in its own right. But through my experience I learned that the hospice was just as much about living as dying. The two processes were inextricable, and the hospice philosophy accommodated this interconnectedness.

One muggy July afternoon I visited Cathy with Monica, my favorite of the hospice nurses. Everything about Cathy pointed to the fact that she was dying. But to Monica, Cathy was not dying. She was living the dying process. Monica found even small ways to encourage and validate the life in Cathy while still acknowledging that death was not far. During that particular visit with Monica and Cathy, the conversation fell to Ensure, a notoriously unpalatable nutrient drink. Cathy said that all this talk about food had made her hungry, and perhaps she would like a small glass of strawberry Ensure. Monica went to the kitchen and returned with the Ensure served in a wineglass with some ice floating in the center.

"It always tastes better if it looks good," Monica said.

"Ooh, how did you know? You picked the perfect glass," Cathy said. "This is beautiful!"

We left Cathy that afternoon sipping strawberry Ensure from a wineglass. Monica had transformed the most basic reminder of illness and degeneration, the inability to eat enough to sustain life, into a life-affirming ritual.

I felt I was a different person from the one I'd been before this summer experience. I had seen what life could hold: Martha, who had managed to move out of the projects and into her own home, died just a year later of metastatic cancer; Maria, the twenty-nine-year-old single mother dying of lung cancer, lay in her darkened apartment with her emaciated mother, who was also dying of cancer. I saw patients who had alienated their families and who, when they were unable to effect reconciliations, died alone. I saw those who were ready to go and those who died with tears streaming down their cheeks.

I loved talking with patients as they wrestled with their mortality and struggled to find closure and reconciliation in their lives. I treasured being invited into patients' homes, affording me the privilege of entering their personal domains and removing us from the sterile formality of the hospital. I valued the humanity in hospice care, and I felt comfortable in that system. After this experience I hoped to devote my career to hospice and palliative care.

As a student I never saw death. I saw dying. But from this brief encounter with dying, I felt I understood better the value of life.

SECOND YEAR

———

Healing Touch

It was a bright September afternoon, and the warm light streamed through the window, lending a soft white glow to the hospital room. Tracy sat in her wheelchair in the middle of the room, and seven of my classmates and I crowded into the small space on the other side of her, hugging the far wall. Her husband was sitting in a chair at the side of the room, but he moved to sit on the bed, closer to Tracy, when we came in. He was a big, athletic man with short dark blond hair and a thick mustache, and he watched over Tracy protectively. She had clear blue eyes outlined by a narrow band of navy blue that had followed each one of us as we came into her room. Thick plastic oxygen tubing stretched from the wall to her neck and fitted over her white tracheostomy tube, a breathing tube inserted directly into her neck.

Tracy suffered from amyotrophic lateral sclerosis, better known

as Lou Gehrig's disease, a degeneration of the nerves causing progressive paralysis and early death. At thirty-five, she was already
in the final stages of the disease. ALS had ravaged her family;
two older brothers and a younger sister also suffered from the
disease.

Tracy was nearly completely paralyzed, and she could no longer
talk. She retained some movement in her left index finger. Her
husband placed her finger in his palm, and she slowly traced
different letters in his hand. When she no longer had strength to
draw the letters with her hand, she spoke with her eyes. Only her
blue eyes remained mobile in her strangely still body. She com-
municated by a letter board similar to the Ouija boards I played
with as a child. Her husband pointed to different letters, and she
blinked when he found the right one. Together they painstakingly
spelled out her words.

Tracy and her husband expected us, and her husband had
brought a video taken of Tracy and her family just a year before.
On the TV screen we saw Tracy and her husband sitting on their
porch at home in the golden glow of the late-summer sun. Tracy
was weak but still able to talk and walk. Dressed in her own
clothes, a plaid blouse and navy blue chinos, she hardly looked
like a patient at all. She smiled and laughed with their two-year-
old son as he played in her lap.

This was the first week of Patient-Doctor II, devoted to learning the physical exam. Our neurologist preceptor brought my
group of eight to see Tracy so he could demonstrate a neurologic
exam. The preceptor seemed to be in his late sixties or early
seventies. He wore a red polka-dot bow tie and blue blazer.

"Ellen," he said to me, "please test the strength of her quadriceps." I was so intent on watching Tracy that I was startled to
hear my name.

Tracy followed me with her blue eyes as I took a step toward
her and bent down to reach her shin. I placed the palm of my
right hand against her pale shin. It was cool and greased with
some type of cream. "Can you kick your shin out toward me?" I
asked her.

Tracy regarded me with tolerant but sad blue eyes. I couldn't feel even the slightest flicker of movement beneath my palm. I looked back toward Tracy. Her eyes widened slightly as if to say, "I tried. I can't."

I took my hand away and stood again. My palm was sticky from the cream, but I was afraid to wipe it off. I looked at my preceptor. "Nothing," I said.

Tracy was the first patient I had ever touched. Already accustomed to interviewing, I had spoken with countless patients at the bedside during my first year of medical school. But I had never touched. I can still feel the stillness of Tracy's leg in my palm; the sad silence in that touch reverberated through my body.

Tracy introduced me to the power and intensity of the clinical touch but also reinforced the futility and failure. The stillness beneath my palm reminded me of the impotence of my skills as a caregiver and of the boundaries of medicine.

It took me quite a while to get used to touching my patients. Touch was so personal and invasive. While my hands could comfort, they now had the ability and even the obligation to hurt as I searched my patients' bodies for clues to the diseases ravaging within. As a second-year student I still had no ongoing responsibilities for patient care. The exam was solely about my touch, and the stories I heard were destined only for my write-up. I felt that I violated these patients, forced a learner's hands on their bodies under the guise of a white coat. But every week I mustered up my courage, marched into a hospital room, and introduced myself to a new patient. I was rarely turned away.

A few weeks after I met Tracy, I found myself in Natalie's room. As I walked into her darkened room wearing my crisp white coat, she hastily pulled her blond hair into a loose ponytail and turned on the bright fluorescent light over her bed. She blinked for a moment as her eyes adjusted to the bright light, and still squinting at me, she smiled and rearranged her forest green silk nightshirt as she sat up in bed. I explained that I was a second-year medical student coming to take a history and perform a quick physical exam, if that was okay with her. She quickly agreed. At thirty-five, Natalie had breast cancer, and she had undergone a mastectomy of her left breast with surgical reconstruction.

"I didn't find my lump. My boyfriend did. We were having kind of rough sex one afternoon, you know, and he pulled on my breast, and I was like ouch! It hurt so bad that I couldn't wear a bra for three days. Afterward we looked for what had hurt me so much, and there was this little lump. It was hard and round, kind of like a marble in my breast. I had never noticed it before. So I made an appointment with my doctor, and she came in and felt my breast. She looked kind of serious like, and she left and brought back someone else to feel my breast. They both seemed pretty worried, and right on the spot they stuck in a small needle and did a biopsy, and it came back positive for cancer. That was about two weeks or three weeks ago. So here I am with a mastectomy. I was afraid to tell my boyfriend that I was going to lose my breast. But he was really nice about it. He said, 'Well, I guess the other one will just have to get all the attention.'

"Do you want to see my new breast?" she asked me.

Before I could say anything, she slipped her silk nightshirt off her shoulders. I was a little taken aback, still unaccustomed to looking and touching. Her new breast was beautiful. It hung softly, looking exactly like her natural breast. Yet it was marred by thick black stitches that reached from her armpit and curved down around the breast, tracing a circle in the center that delineated a smooth circle of pink skin where the nipple should have been.

"You can touch it," she said.

As I slid my fingers slowly across her chest, I felt the gentle rise of the new breast. No ridge of tissue identified this new breast as foreign or fake. I passed my fingers lightly down the breast, careful to stay far away from the black sutures and fearful of causing her pain.

"They're going to do the nipple later," she said. "Do you know how they make it?"

I had no idea. I had never heard of the procedure she had already undergone to create the new breast, much less the further cosmetic procedures she required.

Natalie raised her left arm. "They took out a bunch of lymph nodes here. Will that make my arm swell in the future?"

I didn't know the answer to this question either. "These are important questions. You should write them down and ask your doctor," was all I could offer.

As I delved deeper into the patient-doctor relationship during my second year, the intimacy and power discrepancies seemed more pronounced. Not only was I required to discuss sensitive topics, but now I had to touch my patients as well. I knew I couldn't physically harm someone with words, but all of a sudden my probing hands seemed to acquire ominous potential. While I had grown confident in interviewing patients by the end of first year, learning the physical exam introduced a new, thick layer of insecurity. Even as I gained more clinical skills week by week in my second year, this heightened insecurity reinforced my feelings of incompetence and masked my sense of achievement.

But at the same time, now that I had begun to learn the physical exam, I believed I was on the brink of accomplishing what I had come to medical school to do: patient care. Now I would begin learning about diseases and how to recognize them. During first year I laid the foundation. This year, I thought, I would become a doctor.

My day-to-day routine, especially for the first half of the year, was similar to first year. My classmates and I continued to meet one afternoon per week for Patient-Doctor, but the sessions stretched to four and five hours rather than the two we had spent in the first year. Rotating lecturers taught us how to test the different organ systems through the physical exam, and after the lectures my classmates and I spent time practicing these techniques on one another or on patients.

We took an introductory psychiatry course in addition to our

Patient-Doctor session. Our preceptors brought in patients, and we usually interviewed them as a group with a psychiatrist carefully supervising us. For many of us, including me, it was our first experience with mentally ill people.

We continued to have our regular classes as well. While the first year was devoted to studying normal physiology, we spent the second year studying diseases. The year was organized according to organ systems, and we started with neurology, working our way through the body until we finished with hormones and metabolism in the spring.

At the beginning of the year Kate and I, like most of our classmates, moved out of Vandy and into apartments in the surrounding neighborhoods. Kate and I shared a three-bedroom basement apartment in Brookline with a first-year student. The apartment was old with an exposed brick wall in the living room. My bedroom was small and circular with a fake fireplace and windows that looked out onto a brick wall. The farthest room from the furnace, it was chronically freezing in the winter. But I loved it. I had never lived in my own apartment before, and I appreciated having my own noninstitutional bathroom and my own kitchen. Kate loved to cook, and our house often smelled of fresh bread and stir-fry. Our apartment was a twenty-minute walk from the medical school, and I treasured the escape from the medical area dominated by gray hospital buildings and ambulance sirens. I joined a gym around the corner from the apartment and for the first time in many years worked out under pleasant conditions with up-to-date equipment. It almost made me wonder how I could have tolerated first year.

Roy moved into an apartment just a few blocks from mine with three other classmates. Carlos's apartment was one block from Roy's and a seven-minute walk from mine. Since our apartments were close, we still collected to socialize often. Rather than going out to dinner, we often had impromptu potluck dinners and informal parties. We still gathered religiously every Thursday night to watch *ER*. But the day-to-day intensity of dorm life was gone. We no longer had tea every night or shared our meals every day. For the first time it was possible to carve out more of a personal life separate from the medical school.

While Carlos and I had been good friends during first year, now we hung out more than ever. We played squash and rented videos, and he always walked me home from wherever we had gathered for *ER*. One summery afternoon in late September a group of us went to a nearby park. As we sat near a pond watching a heron pluck golden fish from the murky brown water, Carlos quietly put his hand on my back. The moment was electric. Our friends had already begun wondering aloud whether something was going on between us.

A few weeks later, as Carlos was walking me home from *ER*, he broached the subject. "So, Ellen, what's up with us?" he asked.

I thought he was trying to tell me to back off, so I immediately denied any romantic tension. Fortunately he pressed a little further, and by the time we reached my apartment we had decided to start dating.

I was worried about dating Carlos. Not only were we in the same class, but we also shared the same group of friends. How would our friends view us as a couple? What would happen if we broke up? There was potential for true awkwardness. But I was excited about Carlos, and dating him felt right. It probably wasn't the most practical decision I had ever made, but any concerns quickly slipped by the wayside. Those first months together were magical.

By the second half of second year, socializing had dropped off as our schedules became much busier. We even started to miss the occasional *ER* episode. Patient-Doctor expanded to take up two full days per week, and in late January our class also prepared the Second-Year Show, a musical traditionally written and produced by the second-year class spoofing our professors, our courses, and ourselves. The show demanded endless evenings of rehearsal time.

After the show was over, we began studying for part one of the national medical licensing examination, better known as the Boards, held in June. An extensive multiple-choice exam spanning two days and covering seven subjects, it instilled fear into every medical student's heart. I, for one, had dreaded the exam since I first heard about it when I arrived at medical school. We were required to pass this exam and part two, taken at the beginning of the fourth year of medical school, in order to graduate. We would take the third and final part of this exam at the end of the first year of residency. We gradually drifted apart as we cloistered ourselves to improve our studying efficiency. By the final few weeks before the exam, socializing virtually ground to a halt as we spent endless hours cramming each day.

First Exam

My fingers fumbled to find the pulse of the brachial artery, and I struggled to arrange the blood pressure cuff, my stethoscope, and my patient's arm. "Man, this is the longest setup I've ever had," complained my patient.

"It's my first time," I stammered in reply as I narrowly avoided dropping my blood pressure cuff off the edge of the bed.

Patient-Doctor II was our transition to the clinical world of the hospital. We built on our previous year's experience obtaining a patient's history of illness and experience of disease and incorporated medical observation and diagnosis. We learned to distill a conversation into a medically relevant bullet and to translate a patient's symptoms into clinically important details that pointed to a diagnosis. We learned to tell a clinical story.

With the introduction of the physical exam, the focus of our patient interactions shifted dramatically. It was no longer enough to understand the struggles of being sick and the frustrations of being hospitalized; now we had to determine the patients' cranial nerve deficits and sensory impairments. First year we were expected to extrapolate the human experience of illness from our patients' stories. We were expected to transform patients into people. But second year we learned to transform people into patients.

Masha and I were assigned to be partners for Patient-Doctor that year. Masha was close friends with Andrea, and she exhibited the same emphatic speech and the same urgency about her interests. She was thin, with a narrow figure, and she favored dramatic, trendy clothes.

As Masha and I walked into our patient's room, I wore the same white coat I had worn the year before, slightly less crisp but still brilliantly white in comparison to the more worn coats of the residents working on the floor. This year I also carried a black camera bag holding my new medical equipment. Grasping the pouch at my hip protectively with my right hand, I was acutely aware of the new weight on my shoulder.

This disgruntled patient was my first clinical experience. Masha and I were expected to take a medical history and perform a complete neurologic examination. What began as an awkward moment that afternoon rapidly degenerated into incredible frustration. Our patient was a thin, wizened man, although only sixty-five years old. He lay flat on his bed, the blankets barely rising as they spread over his gaunt form. One wrinkled hand with its network of green, knotted veins rested on his stomach. His mouth was slightly opened as he slept, and his cheeks caved in to fill the emptiness of the open mouth. They puffed out slightly each time he exhaled. His fuzzy gray-white eyebrows twitched in his sleep.

Our patient was in the bed closest to the door. The beige curtain was drawn, and we couldn't see the patient in the other bed. Masha and I pulled up two chairs at the side of his bed, scraping them noisily across the floor, hoping that our patient would spontaneously wake up. Our preceptor, Tim, a fourth-year medical student, stood against the wall to observe our history and physical and offer moral support. Later we would present this patient to our physician preceptor.

"Wake him up!" Tim called to us.

I quietly called the patient's name and gingerly shook his shoulder. He woke easily, fixing his sunken icy blue eyes on us.

"They're second-year medical students here to talk to you and do a physical," Tim told him, breaking the awkward pause when neither Masha nor I could think of what to say. Our patient didn't complain. But during the interview he appeared somewhat confused and repeatedly fell asleep mid-sentence. His icy eyes rolled back in their sockets, leaving only the whites of his eyes still visible, and his eyelids fluttered over the blank eyes.

Partway through the interview I noticed that Tim had become preoccupied by the other side of the beige curtain. The next time I looked back, Tim was gone, but other people came in and out of the room. The patient in the next bed had managed to rip out of his four-point restraints and began yelling. In the midst of this chaos our patient woke up, looked at us, and said, "Can you see the mountain ranges in front of me now?"

Masha and I simultaneously turned around to see if there was a poster of mountains on the wall behind us, but we saw only the standard clock, calendar, and charts pinned on the corkboard panel. Our patient discussed the difficulties of being hospitalized. Then Masha interrupted. "A few minutes ago you said you were seeing mountain ranges," she said. "Can you still see mountain ranges?"

"Shit, man," he said. "I'm not crazy. It was a figure of speech!" I wanted to melt into the floor.

In Patient-Doctor I, where we needed only to talk to the patients to discover their experience of illness, this would have been a notable anecdote for a history composed of any semidirected stories the patient chose to tell. But the responsibility to create a clinical story added urgency. I felt compelled to glean the relevant medical history from this patient's incoherent musings and metaphors. I also knew that at least for now, the physician preceptor

who listened to my clinical summary already knew "the answer."
So I had better come up with the right story.

Our neurologic exam was no better. Intermittently during the
exam the patient's eyes rolled back into his head and he drifted
off into sleep, making it nearly impossible to ascertain whether
he had appropriate eye movements. We still had half the sensory
exam, reflexes, and cerebellar functioning to test when he said,
"Are you almost finished yet? I really have to urinate. Bad." We
took our cue and used the opportunity to end the interview.

"It just can't get much worse than that," Tim told us afterward.
He then demonstrated damage control for presenting this type of
patient. "First of all, you should say, 'Patient was a poor historian,'
to prepare everyone in advance. That way they can't fault you for
a poor medical history and incomplete exam," he said. He also
suggested that we throw in a "bilateral," the term meaning "both
right and left," every once in a while during the review of the
physical exam. It just sounded good, he told us.

Many of our patients recognized the importance of our clinical
story for us. They were asked to participate in our course before
we arrived, and they knew we were expected to discover the right
symptoms and abnormalities and elicit the relevant details of their
histories to uncover the diagnoses that they and our supervisors
already knew. They knew we would be tested on the clinical story
we created. So some of them tried to help us along and smooth
over the early mishaps and foibles in order to save face in front
of our supervisors.

Roy and Carlos were assigned to the same Patient-Doctor session in a different hospital from mine. Both sites followed the
same curriculum, and our experiences were remarkably similar.
They also began by learning the neurologic exam. In his first
patient physical Roy forgot to test the facial nerve by having his
patient squint his eyes and grimace. When his preceptor entered
the patient's room to review Roy's exam, the preceptor asked Roy
about the test for the facial nerve, demonstrating it on himself.
Realizing that Roy had forgotten to test him, the patient inter-

jected, "No, no. Don't you remember? You tested it!" The patient squinted. "It was normal, remember?"

Not fooled for a minute, Roy's preceptor told the patient, "It's okay. You don't have to cover for him."

Carlos was told in confidence by his patient, "Just skip to the head stuff. That's where everything is wrong."

The change in focus from first to second year seemed dramatic. In one of the first Patient-Doctor sessions the two physicians in charge of our course demonstrated how to take a more directed medical history than we had learned in Patient-Doctor I. They sat in front of our group of forty students, and one pretended to be a patient while the other moderated as we asked the "patient" questions. In answering a question about her symptoms, our "patient" began to digress. "I'm really concerned about my son-in-law who was laid off three months ago and can't find a job. So I am responsible for all their child care, and money is tight. . . ."

After a few minutes our moderator turned to us. "Isn't anyone going to stop her?" she said. "In the wards you are under time pressure." Now it was the clinical story that was most important.

As we sat talking one evening, Carlos lamented this transition in our patient experiences. "Last week I saw this patient, and he would have been a great Patient-Doctor One interview. He had a really interesting story. But now I almost felt like his stories got in the way of getting the history and getting to the physical exam," he said.

As eager as I had been to get to the physical exam, I was surprised to miss the freedom of first year to explore whichever issues I found most compelling. I even resented the pressure to find the "right" diagnosis-oriented story. While I knew I hadn't lost the ability to hear the human experience of my patients, I had to learn to channel it toward the medical diagnosis. As I donned my white coat and entered the hospitals again that year, I felt I was one step more clinical.

Naming

When I become a physician, I will determine the diagnoses for many people with many problems. In so doing, I will provide the vocabulary for these people to describe what has gone wrong in their bodies. I will offer the words to explain how their bodies have outwitted them. I will give them a name. Through that name I will link them to other people whose bodies and minds have undergone a similar pathologic process. With the ability to bestow that name, I will carry the power to provide both identity and community. No longer alone in their suffering, patients learn through a name that others have experienced similar difficulties.

Sally came to speak to my classmates and me as part of our psychiatry curriculum. She suffered from insomnia, panic attacks, and intense fear. She was a small, thin woman, and her taut body exuded a restrained energy. The cords of her muscles stood out from her neck as she spoke, making the hollow at the base of her neck all that much deeper. For two years Sally's symptoms were so severe that she quit her job and received disability. She had suffered for almost forty years before a nightmare triggered memories of childhood sexual abuse and she was finally diagnosed with posttraumatic stress disorder. Being diagnosed was an intense relief for her.

"All those years I thought I was just crazy. My whole family used to call me the crazy one. Once my brother called my mom and asked, 'So, how's my crazy sister?' And all of a sudden I wasn't crazy anymore. It had a name. It had a real reason. I could finally understand why I felt the way I did," she said.

Barbara came to speak to my psychiatry class several weeks after Sally. She too remarked on the relief of a diagnosis. Now in her mid-sixties, Barbara had first seen a psychiatrist at the age of ten and had been in and out of hospitals for the past fifty years. Initially her doctors could not provide a diagnosis. In her twenties she was labeled schizophrenic. When that did not seem to fit, doctors changed her diagnosis to posttraumatic stress disorder. She had been given the diagnosis of borderline personality disorder only five years earlier.

Barbara was an extremely bright woman. She was a gifted writer and mathematician who had published numerous articles despite her psychiatric difficulties. But even with these accomplishments she had struggled to find an identity for herself over the years.

Barbara's gray hair was layered in a generic cut. As she spoke in her soft voice, her dark brown eyes remained fixed on the maize institutional carpet. She wore pink sweatpants and a white sweatshirt decorated with a pair of large pastel bunnies. A few keys hung from a beaded metal chain she wore around her neck.

"Because I didn't have a diagnosis, I didn't even have an identity as a mental patient. I didn't know what I was," Barbara told us. Her current diagnosis came as a relief to her. She pointed to the keys hanging around her neck. "Most mental patients wear their keys like this. Now I wear my keys like this to show that I am a mental patient too," she said.

Painful experiences and physical symptoms can be terrifying without the language to describe them. We spent one of our psychiatry sessions on the inpatient pediatric psychiatric ward. The nurse let our group in through the locked door. We saw gaily painted stickers and signs adorning the walls, and toys were strewn across the floor. Two children colored at a miniature yellow table in the playroom. A boy, probably seven years old, repeatedly bashed his yellow sword into the wall just outside his room. A woman crouched down, talking to him. She urged him to stop his destructive behavior and suggested that he might need a time-out break. We peeked through a window in a door and saw a three-year-old boy asleep on the floor of the time-out room, a small room paneled with purple mats like the ones I had played on in elementary school gym classes.

"Hey, Renu!" Renu turned to see a thin black boy wearing blue sweatpants and a T-shirt. A smaller black boy stood behind him. "Hey, Renu!" he said again.

"What are you guys doing here?" she asked, surprised. She stayed behind to talk to the two boys as our group trooped on ahead through our tour of the facility. When she caught up to us, she said, "I know those kids. They play basketball on my street all the time. I see them every day. I just can't believe they're here."

After the tour we regrouped in the conference room. One of the child psychiatrists came to talk to us about the children. He described the case of a four-year-old boy who was severely delayed in his language acquisition. When he finally began to speak, his language was violent. The psychiatrist passed around the four-page list the child's mother had made of his language: "OK to kill Mommy? OK to poke out Mommy's eyeballs? OK to eat Mommy? OK to crush Mommy?"

After several more sessions with the child and his mother, the child drew a picture in which he depicted a situation with a

physician. The psychiatrist passed around a page of scribbled circles and stick people.

The psychiatrist eventually learned from the mother that the
boy had been born without an opening for the anus. During normal fetal development the anus is initially closed by a tissue
membrane, which then dissolves to form an open canal. In this
child the tissue had failed to dissolve, and he was born with a
closed anus. As a result, the child underwent excruciatingly painful monthly anal dilatation procedures to create an adequate opening. These experiences traumatized the boy, who didn't have the
words to understand why he needed the procedures or to voice
his own fears and needs.

Yet despite the relief of finding a name and the identity it implied, a diagnosis could be a mixed blessing. Barbara struggled with
the social stigma that borderline personality disorder carried.
Symptoms of this personality disorder included instability of mood,
inappropriate anger, recurrent suicidal threats, and unstable interpersonal relationships. Our preceptor warned us beforehand not to
ask her about her diagnosis unless she specifically mentioned it.

"I feel like I get treated badly or ignored by hospital staff when
they find out I'm a borderline. I don't like it to be written on my
charts where other people can see it," she said.

With the clinical emphasis in Patient-Doctor during second
year, I became caught up in directing a history and physical to
elucidate the diagnosis. In case after case during our tutorials, I
worked together with my classmates to find the right diagnosis.
While our instructors in our tutorials and in Patient-Doctor
stressed the intellectual process essential to arriving at the correct
diagnosis, my classmates and I knew that ultimately it was the
name that mattered. In the hospitals we would need to find the
right diagnosis to initiate the appropriate treatment to cure our
patients. To us, this was what it meant to be a good doctor: to
make the elusive diagnosis and cure the patient. Yet after hearing
the experiences of these patients, I realized that in the intense
energy I focused on finding the right diagnosis, I had forgotten
the power of a name.

A Conflict of Values

She sat primly upright with crossed ankles, clasping her white vinyl purse in her lap. Her face was carefully made up, and her graying hair pulled into a tight bun. Our entire class had collected in our regular lecture hall for a centralized final exam for our psychiatry course, and this woman's oversize video image dominated the darkened room.

We watched as the psychiatrist on the videotape explained that their conversation would be replayed for Harvard medical students. We saw her agree to be taped. Then we watched as she complained of FBI agents who plotted to kill her, told how her son's penis and testicles had been removed and replaced with a plastic prosthesis, and described how "old men physicians" had performed a hysterectomy on her "with joy and happiness and glee." At the end of the interview she said, "Well, at least I'm not paranoid!" We burst out laughing.

As I wrote down "paranoid schizophrenia" under "Axis I Diagnosis," I felt uncomfortable. I too had laughed at some of the delusions of this woman. The distance provided by the videotape had given us leave to laugh and break the tension of our exam situation. I had observed her actions not only with clinical interest but also with a sense of voyeurism. As I wrote my mental status evaluation, I wondered how this woman, so out of touch with reality, could have understood the function of her interview. How could she have understood how she would be perceived by us? How could I accept her willingness to be interviewed as informed consent? I wondered if in the effort to preserve her autonomy, we had actually violated her rights as a patient and her values as a human being. I wondered if we had failed the patient-doctor relationship by satisfying our needs at the expense of her dignity.

In my first year and a half of medical school, I confronted situations in which my values and objectives conflicted with the patients' beliefs. I tried to understand how best to preserve the patients' values and protect their autonomy. As the medical relationship has become increasingly patient-directed in the past decade, the challenge has become more pronounced. How far and how easily will I compromise my values to preserve those of my patients?

My college years had been a time of self-exploration and personal growth. I tested my boundaries, determined my limits. I created my own set of values. Now that I was in medical school, it was no longer enough to have my own set of values. I had to learn to mesh my values with my patients' beliefs and to validate their ideals without compromising my own. In college I had tried to discover who I was. Now I tried to understand how I would be in the context of who other people were.

My classmate Jonathan found it particularly difficult to contend with issues of homosexuality and AIDS. A very dedicated student, Jonathan applied a military work ethic to studying medicine. There was an intensity about him, although he spoke slowly with a hint of a southern twang in his voice. He had grown up in a religious, conservative family and had gone through military training after high school. In the small town where he was raised, he had had no exposure to openly gay people. His religious teachings could not tolerate the gay lifestyle, and he incorporated this principle into his values. Before coming to medical school, he had been able to avoid issues of homosexuality. But when a gay man with AIDS came to speak to our entire class for Patient-Doctor, he was forced to confront these issues and to begin to decide how he would deal with these people in a medical relationship.

One November afternoon Randy came to our lecture hall to speak to us about his experiences as a gay man with AIDS. Randy was in his early thirties, young and energetic. A tall, thin man, he wore a plaid flannel shirt over a T-shirt and black pants. His face was distorted by many small, flesh-colored lumps, the result of a viral skin infection, molluscum contagiosum. But aside from the disfiguring skin infection, he was as yet otherwise healthy. His partner, however, was in the end stages of AIDS, and Randy desperately tried to maintain hope as his ailing partner grew sicker. During the discussion Randy lamented their diminished sexual intercourse, but he also talked about how they managed to preserve their sexuality and sexual expression in face of this disease.

After the central session we returned to our small groups to discuss Randy's talk. Jonathan was appalled by how explicit Randy had been. "That was a lot more than I wanted to know," he told us later. He had no desire to hear about their sexuality. Yet these details were important to his understanding as a physician how AIDS had impacted on their lives and how to educate them on precautions. Medicine did not give us leave to subject our patients to our moral scrutiny. We had to learn to put aside our prejudices.

While Randy raised a set of issues concerning care for homosexuals with AIDS that Jonathan had not confronted, he had gained some exposure to alternative lifestyles since beginning medical school. He recognized his need to come to terms with his conflicts to achieve better patient relationships. "Before I had no access, no *framework* on which to base my beliefs on this issue.

The exposure of medical school is vital. Now I will always have this experience to look back on," he said after Randy's session.

In my first year, to pursue my interest in medical ethics, I obtained permission to observe one of the local hospitals' monthly clinical ethics committee meetings. Lawyers, administrators, physicians, and nurses sat on the committee. A few members of the committee helped physicians and families make decisions on a daily basis, and they brought any cases with broader themes to the committee for group discussion to develop general hospital policies. Over several months the committee wrestled with issues raised by a patient who demanded that no Catholics touch him in the course of his medical care. The patient had arrived at the hospital for a scheduled outpatient surgical procedure, and at that time he had made his demand. Flustered by the request, the hospital had staffed the operating room with physicians and nurses with Jewish-sounding last names in an effort to respect his wishes. As an equal opportunity employer the hospital administration could not ask physicians and staff about their religious affiliations. To complicate the matter further, after the procedure the physicians caring for the patient discovered he was a paranoid schizophrenic. They were outraged that his request had been honored.

While the case of the schizophrenic seemed easy to dismiss as violating physicians' rights and values, in some situations a patient could legitimately ask to select medical staff by gender or race. What if the person had been raped or beaten? Alternatively, the patient could be sexist or racist. Who should decide whether the request was valid and what criteria should be used? The patient-doctor relationship was very sensitive and invasive, prying into the most private and even frightening parts of patients' lives as they faced physical or mental crises. Should we therefore err on the side of honoring patients' needs over those of the medical staff to whom these relationships are more routine? This issue put before the committee formalized the struggle as it attempted to mesh conflicting values into a unified vision of medical care.

The anger of the physicians took precedence in determining the policy based on the case of the schizophrenic. The doctors felt the need to be protected from discrimination, and the committee created a policy to limit patient requests to select physicians by gender, ethnicity, and religion. I felt dissatisfied by this position, fearing that because this was such a straightforward case of unjust discrimination, members of the committee had been unable to step back far enough to consider the needs of the patients. The patient-doctor relationship was more foreign to our patients than to us, and I thought we needed to protect their sensitivities.

Complicating the struggle to achieve this balance between patient and caregiver was the fact that many of the patients I met through Patient-Doctor were in crisis. In the face of illness, they reevaluated their principles to accommodate their new situation. Thus, while I tried to maintain my own sense of self-definition within a patient relationship, the patients themselves were in flux; this created a dynamic and precarious balance.

I thought back to my patient John and his wife, Suzanne, whom I had met last spring as part of Patient-Doctor I. Although they both were Jehovah's Witnesses, John had just received a heart transplant after his own heart had been destroyed by a virus. Organ transplantation was prohibited in their belief system. While Suzanne respected his decision to accept the heart, she herself would have been unwilling to risk the afterlife repercussions of organ transplantation.

"I couldn't help it," John said. "In that moment I just wanted to live."

This need to achieve a compromise between conflicting values was a new challenge, one that I had only just begun to recognize. While I merely observed these situations, I could already envision conflicts arising when I would be the caregiver. I did not yet have any idea of how I would achieve resolution or, indeed, if it was even possible.

The Pelvic

I did my first pelvic exam on Valentine's Day. And my first prostate exam. And my first testicular exam.

The genital exam was taught in an evening session where my classmates and I practiced on professional patients trained to teach us the necessary skills—two hours for the female exam and another two for the male. As February 14 approached, I was filled with anxiety. This exam was by far my least favorite part of my own regular physicals, and I was unsure how I would respond now that I had assumed the physician's role.

I had spent the last one and a half years of our Patient-Doctor clinical skills course learning how to question patients about their sexual practices and concerns. I worked to become more comfortable with the issues we probed and the information we gleaned. I struggled to make my patients—and myself—feel comfortable discussing topics fraught with taboo. But now I crossed the physical boundary. It was no longer discussion in the abstract; I observed, examined, and palpated. My classmates also struggled with these issues. For the first time we had to extricate ourselves from our sexual associations, cultural values, and personal beliefs about genitals and transform them into a purely clinical experience.

After watching a very clinical—and very graphic—video of the pelvic exam and practicing the exams on plastic male and female dummies in preparation for our teaching session, my classmate Scott said, "Will sex ever be the same for us again?" As we struggled to achieve clinical objectivity, sometimes the boundary felt blurred.

Our first exams were staggered to accommodate small student groups; they had been running nightly sessions for months by the time my turn came. Those of us scheduled for later slots questioned classmates who had already gone through their training session. It really wasn't too bad, Masha reassured me. "You get so caught up in looking for the structures, you forget what you're actually looking at," she said. Her "patient" had bruises on her abdomen from students attempting a bimanual palpation of the ovaries. In focusing on the physical, some of my classmates lost sight of the person.

Andrea was thrilled with her first pelvic exam. She came into tutorial the following morning glowing from her experience. "I'm going to be a vagina doctor. I just know it. I'm going to be a *vagina* doctor!"

Despite my trepidation about the upcoming exam, Scott said he wasn't particularly concerned about the session. These patients were trained. They were knowledgeable about the exam, comfortable with being examined, and fully aware of our complete inexperience. "The first exam doesn't bother me," he said. "It's the second that I'm worried about." He feared his first exam on a real patient.

When confronted with real patients, my classmates and I would be forced to grapple with our patients' perceptions of the exam and their issues concerning genitals, not to mention our own. Insecurities about our clinical inexperience would compound the struggle.

Patients are legitimately concerned that we physicians will fail in our attempt to extricate the sexual aspect of genitals from the clinical exam. In particular, women have become increasingly aware of the sexual harassment risks involved in the pelvic and breast exams. They routinely request female examiners, and in our lecture on these exams the physician recommended that we always perform them with a chaperone in the room. Even she, a woman, did not do them alone, unless she had already established a long-term relationship with the patient.

Yet the increasing desire for patients to protect their privacy frustrated some of my male classmates in their attempt to learn the female exam and gain clinical experience. The course description of the obstetrics-gynecology clinical rotation at one of the Harvard sites warned: "Many HCHP [Harvard Community Health Plan] patients strongly prefer a female physician and decline examinations by male students."

A fourth-year student complained, "If you're a guy, you just can't learn it. None of the patients will allow us to watch. If you ask first, they always say no. I didn't do a single gynecological exam and I watched only one birth during my entire ob-gyn rotation. So you have to be aggressive if you want to learn it," he said. Another male student disagreed. But he emphasized the importance of getting to know the patient, if possible, before starting the procedures.

If the boundaries between emotional and clinical, cultural and medical were difficult for us to draw at times, our patients also struggled to make the distinction. When entering the intensely private relationship of a physical exam, they sometimes blurred the distinction between a professional and personal relationship. Our patients flirted.

One of my classmates practiced a mental status exam on a thirty-five-year-old male patient. Clearly psychologically competent, the patient was bored with the simple exercises to determine memory, reasoning, and judgment. When my classmate pointed to her shoe and asked him to identify it, he said, "Nice foot, nice ankle, nice knees, nice thighs. Want me to go any higher?" She was taken aback.

"At the time I just sort of laughed it off. I didn't know what else to do. I sort of thought it was my fault, like maybe I hadn't set a formal enough tone. Maybe I was just too casual and jokey," she said.

An older male patient kept touching Renu's hair and brushing it back as she leaned over him with her ophthalmoscope to visualize the retina on the back surface of his eye. To do the exam correctly, the examiner had to be within inches of the patient. "You can come closer. I really don't mind," he told her.

"The worst part was that he clearly knew that I knew he was flirting with me. He was from another culture, and maybe that was more acceptable where he came from. But it was very uncomfortable. I didn't know what to do," she said.

My male patients often thought I, as a young, small woman, was cute. Some of the older men identified me with their granddaughters, and one patient even surprised me with a familial peck on the cheek as I left. But some of them used a distinctly more sexual tone. I preferred a slightly less formal tone in my patient relationships, but these attitudes made me uncomfortable. I never found a way to discourage this atmosphere from creeping into my patient experiences. I even accepted it because these men were often willing patients and didn't complain about the two-hour histories and physicals I subjected them to.

The struggle to maintain a distinction between the clinical and the personal invaded the classroom as well. As we learned the different components of the physical exam, we practiced on one another. While we did not perform the genital exam on one another, some of the other procedures were nearly as sensitive. As we entered these pseudo patient-doctor relationships with one another, the boundaries between companion, colleague, and patient became blurred. We didn't fit into any category.

Carlos's patient-doctor group learned to palpate the inguinal lymph nodes, which were in the groin. I asked him how the session had been. "Revealing," he said.

Carlos acted as the patient for the group. As the patient he sat in his boxer shorts in front of a mixed-sex group of our classmates as the instructor demonstrated the exam and the students practiced. Both male and female classmates in his group had to feel under his shorts to palpate the nodes. Roy wore double underwear—Jockeys *and* boxers—to minimize the invasiveness and embarrassment inherent in the situation.

After all the buildup for the event, my first internal pelvic exam was not such a big deal. I think one reason for my fear was that as a woman and unlike a man, I had never really seen my own genitalia. This was foreign territory. And since I experienced my own exams as a humiliating procedure, I feared inflicting the same humiliation on another person.

Lisa, our professional patient, was completely at ease. These "patients" were trained to teach us the genital exams, using themselves as models. Because I arrived a little early, I met Lisa outside the examining room just as she was coming in. Lisa was probably only five or six years older than I was. An obese woman, she had permed shoulder-length mousy hair and pale blue eyes. As we waited for the administrator organizing the evening program, I felt compelled to make conversation. "Thank you for doing this for us. We really appreciate your coming to teach us," I told her.

"Oh, yeah, well, I like coming. And it's a great way to make money," she said. Lisa had burned out in her teaching job a few years ago. Looking to change professions, she had come across an advertisement for volunteers to act as patients for medical students. "It just seemed like a great opportunity at the time. And so many women die of breast cancer that I think it's important that you guys know how to do a good exam. I'll probably do this for a few more years before I move on to something else."

When our turn came to examine Lisa, Scott and I entered the room to find her in a johnny sitting on the examining table. Lisa had been trained both to show us how to do a proper exam and to act as the patient. She knew exactly how each part of the exam was supposed to feel and helped modify our technique to improve our skills. First, she carefully described the necessary motions women must make to obtain an adequate visual exam of the breasts. Then she lowered the top of her johnny to reveal her pendulous breasts. She flexed her arms as we watched, and then lifted them and clasped her hands behind her head. We watched her breasts change shape as the chest wall muscles behind them tightened and relaxed. Then she lay back so we could practice the breast exam. Scott went first. "No, you have to press harder. I could still have a lump there that you wouldn't feel."

As his fingers worked meticulously across her breast, he found dense fibrous scar tissue at the base of her breast. "That's from my breast reduction surgery. It's completely normal. But you should feel it so that you know what it's like." As he lifted her breast to reach the tissue better, I noticed the three-inch ruddy crosshatch scar.

Then it was my turn. I carefully moved my fingers in small circles across the entire surface of her breast. "That's good. The pressure is just right." I found the fibrous shelf of scar tissue at the base of her breast, and I ran my fingers across the scar.

Now came the part I dreaded. I arranged the light for Scott as Lisa pulled her johnny up to cover her shoulders and then spread a paper sheet across her lap. She pulled out a blue plastic hand mirror while I ran warm water over the speculum, the gray metal instrument used to open the vagina and visualize the cervix, the opening of the uterus. She showed Scott how to feel her labia and the soft mound of her mons over the pubic bone for masses. She told him how to spread her labia majora to reveal the wrinkled pink labia minora inside.

As Scott inserted the speculum into her vagina, Lisa held the blue plastic mirror so she could see what he was looking at. "Okay, now that the speculum is in, you have to rotate it ninety degrees and push it toward the back of the vagina. And most important, be careful to keep pressure toward the floor on the speculum. You don't want the speculum to ride up and hit my clitoris. That's excruciatingly painful for women."

Scott advanced the speculum. Small beads of sweat accumulated on his brow.

"Oh . . . you need way more pressure. You're getting close to my clitoris," Lisa warned.

When it was my turn, I tried to follow the directions she had given Scott. I inserted the metal speculum and then rotated it ninety degrees.

"Watch my clitoris. A little more pressure."

I pushed the speculum to the end of her vagina and then opened the lips of the speculum. I saw a pink tissue wall with streaks of creamy white mucus. This didn't look at all like the cervix.

Lisa surveyed the situation in her hand mirror. "Why don't you try closing the lips, aiming more toward the floor, and pushing the tip in a little farther?"

I pushed the speculum a little farther and opened the blades again. Still the same pink wall. I could feel my face warm under the hot light of the lamp. I tried one more time, and this time was grateful to see the raised pink doughnut of the cervix with the central red spot identifying the passageway into the body of the uterus.

The bimanual exam was much more difficult for me. I pushed my gloved and lubricated right fingers into her vagina, reaching for the cervix I had just seen. I pressed my left hand into her soft abdomen, trying to press on the uterus and bring the cervix into reach of my finger. I reached with both my hands, but the cervix remained elusive.

"Push your fingers in farther," Lisa said.

I leaned toward her, in part using my weight to force my fingers a few millimeters deeper into her vagina. I couldn't find the uterus in the ample softness of her belly.

"My uterus won't be all the way up there. You've got to press lower. Here." She reached down and rearranged my hand. Finally I was rewarded by what I thought was the light touch of the cervix passing across my finger. Relieved, I finally extricated my fingers from her vagina still a little unsure of what I felt and fairly certain I would not be able to replicate that exam on another patient.

Although I had worried about how I would feel in the room with Lisa, she was so matter-of-fact that she put me at ease. Once reassured that she was comfortable with the exam, I could relax and focus on finding and examining the necessary structures. It was hardest to watch Scott attempt the pelvic exam, knowing that my turn was fast approaching. Once involved in the exam, I became so absorbed in my efforts that I forgot my personal inhibitions. After leaving the room, I thought that while I might be technically incompetent when examining my first patient, I had at least begun to tackle my fear of the exam. I felt confident that I would be able to treat the pelvic as a routine part of the physical exam.

Ironically, I was less afraid of the male genital exam, although I wasn't sure why. Perhaps I had spent so much anxious energy preparing for the female exam that I couldn't get as worked up over the male exam. Or perhaps my experience with Lisa was such a relief that I was no longer worried about learning the male exam.

I recognized our patient's name. Mr. Miller had taught Kate, my roommate, and her partner the genital exam a few weeks previously. He was a tall, thin man in his mid-sixties. He and his wife had trained to be patients a few years ago when his wife was diagnosed with breast cancer after her primary clinician had missed a lump. "We really want you to learn this well so that you won't miss any lumps." His wife was in poor health and had recently stopped acting as a patient.

When Scott and I entered the examining room, Mr. Miller wore a blue pinstripe shirt with blue chinos. We were the second session for the evening; he had already taught two of our class-mates. First he quizzed us on the statistics for testicular cancer and made sure we knew what to look for on the physical exam. Then we talked about language.

"Never ask a man to spread his legs. It's too demeaning," he said. "Instead you should ask him to move his legs apart. If he gets an erection, you can just stop the exam and continue later." We discussed how to teach a man to do his own exam.

After the short talk we left the room so Mr. Miller could change. When we returned, he was wearing a white T-shirt, blue chambray boxers, and black dress socks. We practiced the exam on him, all the while demonstrating how we would teach him to do his own self-exam. This exam was much less involved than the female exam, and it took only half an hour to learn.

I palpated the shaft of his penis and practiced rolling back the foreskin. "Never, never forget to replace the foreskin. It's a very sensitive area, and it can be painful if you don't replace it," Mr. Miller reminded me. I released the foreskin and practiced rolling it back between my thumb and first finger again. Afterward I tried my first prostate exam. I had been told the prostate would feel similar in consistency to the tip of my nose. After following his directions, I inserted my finger into his rectum and felt the smooth firmness of the prostate beneath the pad of my finger. I felt across the smooth base of the prostate, but try as I might, my finger would not reach to the top of his prostate. "You've got to really lean into me." I pushed a little harder, still impossibly far from the apex of his prostate. Finally, I removed my finger. I was relieved to have all my first genital exams over.

Relationships

I always thought getting married would be as simple as falling in love. The challenge was to find the perfect man. My life partner and I would surely overcome the obstacles and surpass the hurdles to forge a life together. Then I came to medical school.

As I watched students ahead of me as well as clinical instructors struggle to manage relationships, families, and medicine, I became acutely aware of the challenges ahead. While I could not yet know all the obstacles, I already envisioned some. Long hours, nights on call, and clinical dilemmas would sap our energy and leave little left over to nourish a spouse or a family. Critically ill patients and medical emergencies would be hard to push aside to devote more time to family. How would we say no to the patients who needed us?

After one of our Patient-Doctor sessions Masha asked our preceptor what time he got home. "I'm here by five or five-thirty every morning, and I often don't leave until nine at night. Sometimes I go for days without seeing my kids. But," he said, pulling a photograph of two toddlers out of his pocket, "I carry a picture with me every day."

In a recent panel discussion on managing medicine and relationships, a member of a happy dual-high-intensity-career couple said, "There is no peace in this life, and there is no peace in this marriage."

That evening we heard the testimony of four couples at different stages of their careers who had managed to preserve their relationships despite the stresses of dual medical careers. They were honest about the challenges to their careers and their families. They wrestled with issues of child care. They fought over whose career should determine where to live and when, or whether, to have children. They struggled to find time to spend together. They battled chronic exhaustion. They were the successful ones. The divorce rate among physicians was rumored to be well over 100 percent when failed second marriages were taken into account. A *New England Journal of Medicine* article offered a more realistic statistic, with psychiatrists having the highest divorce rate at 50 percent and pediatricians with the lowest at just over 20 percent. For each of the couples on the panel, there were many more who failed to survive.

One couple on the panel, who were Harvard Medical School classmates when they married thirty-five years ago, had a more troubled past. The wife described their earliest years as "blindingly unhappy" until they each learned to balance their professional aspirations and medical obligations with the need to nurture the marriage and each other. She chose to spend several years at home with their children; her husband believed this choice had hampered her medical career.

Carlos and I attended the forum together. We had been dating nearly five months now, and I was completely in love. Just before Christmas we spent a weekend in New York City. We spent the days enjoying the Christmas decorations, walking in Central Park, and visiting museums. We lingered over long dinners in the eve-rings. There wasn't a moment of epiphany, but I believed that the weekend had cemented our relationship and heightened the intensity of our commitment. Yet we were panicked thinking about the years ahead. As much as we loved each other, neither of us was certain that we could withstand the pressures of a dual medical career or even of medical training.

After the panel discussion we weren't sure whether to be encouraged or depressed by what the couples said. Each couple had managed to maintain their careers and their relationships, but the stresses of the workplace had impacted strongly on the marriages. While most had eventually found a satisfactory balance, it seemed that all the relationships had only barely survived.

"I was hoping someone would say, 'Don't worry. It won't be as hard as you think.' But nobody is saying that," Carlos said afterward.

During our cardiology course our instructors sponsored a cardiology interest lunch to encourage us to pursue careers in this field. We had the opportunity to question a cardiothoracic surgeon, a cardiopathologist, and two cardiologists about their lifestyles. "You can ask us anything," they said.

The conversation naturally drifted to work schedule and family life. They worked eighty-hour weeks, from five in the morning until nine at night. On the rare afternoons they got out of the hospital early they rushed to the lab. We asked if they were married. The cardiothoracic surgeon was single. The pathologist was married—actually, remarried. "Well, I should tell you that I got divorced during my residency. But it wasn't just the residency," he quickly added. "There were problems in the relationship too."

When faced with single parenthood, he chose cardiopathology as the only cardiology career with a feasible work schedule. The other two cardiologists were married but had chosen not to have children because their jobs were so demanding. "It just wouldn't be fair to the kids," one of them said.

Loan repayment was not designed for families either. In a recent loan counseling session, financial aid counselors distributed a worksheet entitled "How Far Will My Paycheck Go?" The budget was divided into categories, including what we had to pay for and what we wanted to pay for. Any left over was discretionary income. Our spouse's loans were accounted for. Car payments, vacations, and even pet care were budgeted. A certain sum was allotted for retirement. But there was not space to include the expense of having a child. You could own a cat, but you couldn't have a baby.

Quite a few of my classmates had weddings planned for the summer after second year. Unofficial tallies tabulated by classmates during our more boring lectures suggested that 30 percent of our class would be married or engaged by the beginning of third year. Kate, who was one of those getting married that summer, said to me, "Statistically we are not all going to make it. But of all the couples, who would I pick out? Alexa and Andy? Kevin and Anne? What about me? I have to believe that we'll make it."

I always expected that managing medicine, marriage, and family would be difficult. But I never expected it would be this hard. This year, with flexible classroom schedules and regular hours, it had been easy for Carlos and me to build a relationship. But I feared what lay ahead.

The Show

It was Saturday night. As the dusty brown curtain drew to a close for the final time, 150 of us crowded to the back of the stage, spilling into the wings to get out of the way so no one would be caught in front of the curtain. Our communal cheer of exhilaration and relief drowned out the final strains of "YMCA" by the Village People and overwhelmed the audience's applause. We were done.

It was the final night of the Second-Year Show, a musical written and produced each year by the second-year medical students. All in good fun, we mocked our professors, our classmates, and ourselves. Ours was a four-hour epic. When 150 compulsive perfectionists gathered, there was no limit to the time and attention devoted to a project. This performance culminated a creative process, begun seven months earlier, that grew into marathon rehearsals consuming every free moment and dominating every conversation for weeks.

In addition to fulfilling our obligation to tradition, the show gave us the opportunity as a class to identify some of the issues we had struggled with over the past two years and to reflect on the experiences that defined our time at Harvard. This was the last time we would come together as a class before we cloistered ourselves to prepare for the boards, part one, national medical licensing exam in the spring and then entered the hospitals on our disparate rotation schedules. As we prepared to quit classroom academics for the world of the hospital, the show offered a sense of communal closure on the last two years.

Concerns about our overwhelming debt and about the future of American medical care dominated the show. Because of the rapidly changing political situation of health care reform, we had no idea what medicine would be like when we finally began to practice. Many of us, myself included, would be at least one hundred thousand dollars in debt by the end of our four years. My bank had refused me overdraft protection because my debt-to-income ratio was too high. There was tremendous fear that with the decline in physician salaries, we would be unable to pay off our exorbitant loans.

But more than our own personal finances, we worried about the economic reforms in medicine itself. As we watched the health maintenance organizations (HMOs) assume more and more influence, we feared that physicians would lose control to the politicians and businesspeople who would financially streamline medicine. We wondered how we would adapt to all the changes in medical practice.

Given the level of concern, it was not surprising that the plot of the show reflected these fears. The beginning of the show depicted Harvard financially devastated in the world of managed care. The school no longer has funds to provide financial aid for medical students. In the show Tori Spelling of the Fox series *Beverly Hills 90210* applies to HMS. Her TV mogul father, Aaron

Spelling, offers to buy the medical school and turn it into a TV network in exchange for acceptance of his daughter. Dean of Medical Education Daniel Federman agrees to sell out. When Federman suffers a near-fatal accident, the directors decide to kill him off to improve ratings. At the last minute Tori recognizes Dean Federman's genetic disorder and cures him. But because ratings have plummeted, Aaron pulls out of Harvard, leaving HMS financially bereft again. In a deus ex machina finale, one of the renal physiology professors generates enough money to keep HMS running for twenty-five years.

The opening scene heralds the Age of Indebtedness in a remake of "Aquarius" from the musical *Hair*. But what was more frightening than the poverty itself is the power business and government gain as they assume financial control of medicine:

> When the Newt is in the Senate House
> And HMOs are on the rise
> MBAs will mold health care
> And welfare's swift demise
> This is the dawning of the Age of Great Poverty
> Can't you see?
> Poverty

Moreover, in the show medicine changes rapidly, arbitrarily, and unpredictably depending on financial fluctuations. In reality Dean Federman was a respected endocrinologist famous in our class for his signature seersucker suits and bow ties. While he did not play a large role in our day-to-day education, he was instrumental in instituting the New Pathway along with its creator, Daniel Tosteson, the dean of the medical school. But in the show, when Dean Federman succumbs to Aaron Spelling's financial influence, he sacrifices science for glitz and academics for appeal. Similarly, we worried that HMOs would sacrifice good medicine for more money. We wondered how we would function in a financially constrained atmosphere that would seek to dictate our medical practices. The drive for economic reform had already shown its effects on our class. As we learned the importance of taking a good history and performing a thorough physical exam, we wondered how we would accomplish both goals in the ten- to fifteen-minute office visits mandated by HMOs. Whenever a clinician taught us about a new procedure or discussed a technological advance, someone invariably asked how much it cost and whether insurance would pay.

In the show, when Aaron revokes his financial support, HMS and Dean Federman are again without allies. But when HMS Professor Julian Seifter generates enough revenue to rescue medicine and restore its integrity, Dean Federman's allegiances switch again in favor of the new source of funding. In the final line of the show, Billy, a medical student, struggles to keep up with the rapidly and unexpectedly changing face of medicine. "But what am I going to do now?" he asks. "I've wasted all my time studying to be a television doctor. Now how am I supposed to start learning *real* medicine? At this rate, I'm *never* going to be a star on *ER*." We were afraid that like Billy, we would be caught unaware in a medical and political environment we were unprepared for.

Issues of identity and transformation provided a more subtle theme throughout the show. Dean Federman and Aaron announce the merger of medicine and Hollywood at orientation for Tori and the other incoming first-year medical students. Since HMS is now both a school and a network, students have taken on a dual role. In Aaron's welcoming speech, he says, "As part of the DOXX network, you will be embarking on an unprecedented journey into the world of televised medicine. You will become actors even as you become doctors. Superstars as you become subspecialists. And, someday, Oscar winners as you become Nobel laureates." Becoming a doctor felt like a facade, a masquerade. Uncomfortable with our new role, we thought it assumed and artificial.

This theme continued as Dr. Beverly Wu, who in real life was in charge of the first-year Patient-Doctor course, introduces the students to the patient-doctor relationship. "The sanctity . . . of the patient-doctor relationship is symbolized by the white coat the mantle of respect for our profession." Yet what was truly remarkable about the coat "is that underneath it you can hide who you TRULY ARE—your naughty side," she says.

The white coat covered our discomfort with our role. While it identified us as medical, it also disguised us. We were impostors. The refrain to the song "Coat" sung to Madonna's "Vogue" emphasizes this: "Wear the coat/Let patients think you're a doctor [think you're a doctor]/ Hey hey hey/ Put on, coat/ Never let on that you don't know/ You know you can do it."

In another scene HMS/DOXX students, including Beavis and Butt-head, are instructed to assume the power implied by the patient-doctor relationship. "You're God, dammit. YOU ARE GOD! . . . Do you wanna play God?" The students feebly attempt to muster a sense of authority. "So, like, I'm pretty much God," Butt-head says. While this was clearly an exaggerated vision of the power differential, it was a discomfort that resonated with our patient experiences.

In many ways the show represented an ideological boundary between the preclinical classroom years and the clinical hospital experience. There were several more hurdles, such as Boards, to surmount before we ended this half of our medical school experience, but this communal experience cleared the air of the experiences that had accumulated over the previous two years and allowed us to move on to the tasks of the next two. Logistically the end of the show marked the end of leisure as we buckled down and began studying for the Boards exam, now only a few months away. Even though dirty snow still blanketed the streets and made spring seem impossibly far away, we felt our exam rapidly approaching.

Boards

Our instructors told us the exam was benign. We would surely pass, they said, and it wasn't even that important to do well. Third-year students told us to use the exam as an opportunity to pull together the loose ends from the previous two years. Nevertheless, the Boards, part one, a two-day exam covering basic science concepts, struck fear into even the most stalwart of second-year hearts.

This exam defined the moment of transition from the classroom to the wards. It forced us to relearn all the facts and details we had forgotten over the previous two years and then some. It forced us to prove our competency in the principles of medicine and demonstrate that we were ready for the next stage of our education.

Despite the lofty goals of the exam, only the most competitive residencies demanded superb Board scores for admission. For most of us, we needed only to pass. First year, as Roy and I sat in the library studying for one of our pass-fail finals, I had wondered why we worked so hard. "I want to pass with dignity," Roy had told me. But in preparing for this exam, he changed his philosophy. "Now the dignity is in just passing," he said.

Even so, sometimes it seemed that just passing wasn't enough. I, along with many of my classmates, invested this exam with added meaning. In a medical world where we constantly wondered about our roles and doubted our abilities, this exam proved to us that we belonged. It proved that in the New Pathway world of limited requirements we had learned enough. It proved that we would be good doctors. While I knew that no one exam could possibly determine so much, I still struggled not to use this score as the sole marker of my medical ability.

Even the third years, who urged us not to worry too much about the Boards, put in high-intensity, grueling hours in the libraries while preparing for their own exam. One third year, in particular, who spoke on a Boards preparation panel encouraged us not to take this exam so seriously. But I remembered having spoken with that same person the year before, just a few weeks before he had taken his exam. He had proudly told me then how he had moved all his books and a coffeemaker into one conference room in the MEC to expedite his studying efficiency.

Several months before the Boards the medical school administered a practice test. After taking the exam, I worried that I hadn't passed. I knew a failing grade would make me nervous about my ability to pass the real exam, since I had already begun preparing for the June sitting. All week I worried: What if I didn't pass? What if I couldn't pass? Then I got my score. I passed. Still, it wasn't good enough. I hadn't done well. After looking at our scores, Carlos said to me, "Sometimes I feel trapped by my perfectionism."

My classmates and I devised study schedules for ourselves to prepare for the exam in a methodical manner while inducing the minimum anxiety. We all put tremendous energy into determining the number of hours we wanted to spend, how to distribute them among the subjects, and in what order to study them. One third year at our Boardsbuster meeting advocated a strict spreadsheet allocating a certain amount of time to each chapter of each review book. Chapter three in the histology review book was worth only twenty-nine minutes, but he appropriated three hours and fifty-eight minutes for his cardiovascular pathology review. He checked off each box as he completed the given amount of time.

"At first all my friends laughed at my schedule," he said. "But by the end they all wanted a copy."

Despite the careful thought and endless conversation devoted to developing these rigorous study schedules, only a very few of us managed to abide by them. Carlos was the sole person I knew who managed to keep up with his plan. I finally fell so far behind my schedule that I no longer felt stressed about trying to catch up.

But the pressure to spend every spare moment studying was at times overwhelming. One Saturday morning Roy's roommate inadvertently locked himself into his bathroom. One hour later the locksmith and Roy finally pried open the bathroom door to find the roommate sitting fully dressed in the bathtub reading *Neuroanatomy Made Ridiculously Simple;* it was the only review book thin enough to slip under the bathroom door.

Another classmate became progressively more concerned about spending every possible moment studying as the exam drew nearer. She studied everywhere: in the car, while cooking dinner, while walking the dog. Finally, she realized that if she put her notes in a Ziploc bag, she could take them into the shower as well.

Studying for the Boards imposed a great strain on our lives, regardless of whether we studied intensely to earn a top score or were pleased just to pass. One of the third years on the Boardsbuster panel and his girlfriend went away from HMS in the final few weeks before the Boards. They hoped to escape the stressful atmosphere generated by their classmates at school. But even away from school he and his girlfriend struggled with the stress.

"It was a great test of a relationship," he said. "Even more than infidelity."

Preparing for the Boards was stressful. While I heard the advice of my elders, my anxiety made it virtually impossible to heed. I vacillated between confidence and complete panic. Carlos raced to keep up with my rapid cycles, and ultimately his calm and steady presence afforded me a sense of stability.

The Horror, the Horror

In April I watched my first operation. I was prepared. Our Patient-Doctor coordinator warned us that we might faint. "Everybody does the first time," she said. "It's the skin incision that really gets you. Once you get past that, everything else is a piece of cake," she said. My site coordinator was a middle-aged, highly energetic woman who talked fast and moved even faster. Although she said *everyone* passed out, I felt certain she never had.

As the purple and yellow crocuses forced their way through the gray Boston snow, which slowly melted away to reveal the matted, soggy grass beneath, our Patient-Doctor instructors prepared a mini-introduction to third year for us. In addition to our regular routine of practicing exams on patients, reporting the results to our advisers, and writing them into formal notes, my classmates and I spent one morning per week following residents in different subspecialties. These second-year rotations through surgery, pediatrics, cardiology, neurology, and the emergency department (ED) introduced us to the world we would encounter during third year, when we worked in the hospitals.

I was anxious to pass out. Surgery, my first rotation of third year, began on July 1, and I was glad to have the opportunity to feel the gruesomeness of the situation before I entered the more pressured atmosphere of the wards. It would still be pretty embarrassing to pass out now as a second year, but at least I wasn't being graded. I wanted to be desensitized before I arrived on the wards in two months.

So I was somewhat disappointed when my surgery preceptor for the day arranged for a classmate and me to watch a nonsurgical drainage of a pelvic abscess. Although a little stooped with age, the surgeon was wiry, and we raced after him all morning. He wore scrubs and sneakers and a blue scrub cap, all the while chewing gum and spitting on us as he talked. Although the abscess drained enough yellow-brown pus to fill half a can of Coke, it did not have the drama of an open incision. Our surgeon, however, was quite excited.

"You know, team," he said to my classmate and me, "there's nothing so satisfying as pus. I love pus! She'll be uncomfortable for the next few days, but after that she'll feel just great."

The gore of medicine has always frightened me. I cringed at the thought of reducing dislocated joints, stitching bloody wounds, dressing gangrenous toes. *ER* paraded blood and guts on prime time in trauma after trauma. After watching every Thursday, my parents and nonmedical friends told me, "I could never do what you're doing." I wasn't exactly sure I could either. I thought this was the horror of medicine.

Yet after that surgery experience I realized that becoming desensitized was not the challenge I had envisioned. Nearly everyone passed out in his or her first trauma and first surgery, but over time the horrible became commonplace, the visceral mundane.

"The surgery itself gets pretty boring after a while. It's the human interest in the patients that makes this job interesting," said another surgeon, with whom we spent several hours.

Following the initial disappointment of the pelvic abscess, our surgeon took us to watch the reconstruction of a breast after a complete mastectomy. It was the same operation that Natalie, my patient with breast cancer, had had. In the procedure the rectus abdominis muscle was removed from the abdomen to be used as tissue for the new breast. When we entered the operating room, the woman's face was hidden behind a steel blue curtain. Her abdominal muscle had already been removed, and the deep, rectangular crater left behind was packed with gauze. The layers of yellow fat and fascia and pink muscle were visible at the margins. Three surgeons carefully dissected off the top layer of skin from the tissue that used to be her right breast. The lumpy tissue revealed after her skin had been dissected away now sweated delicate rivulets of blood.

When the surgeon working on this woman heard we were medical students, he offered a brief explanation of the procedure. "This is really an excellent procedure. You can see how healthy the tissue is," he said. He scraped his scalpel against the denuded chest wall and demonstrated the ample blood supply by milking a few more drops of blood. "But as you can see, this is quite involved surgery. It's not for everyone. This requires commitment, you see. Commitment. Those are exactly the words I use when I screen patients for this procedure. Commitment. The bonus is that you get a tummy tuck at the same time. So for women who can handle the surgery, it's a great procedure."

The anesthesiologist interrupted to tell the surgeon that the patient's blood pressure had dropped quite low. "Let's give her her first transfusion now," the surgeon said. "I think she's lost most of the blood she's going to lose by now, anyway. Thanks for letting me know."

Now I had seen the desensitization that would rescue me. It didn't seem as hard as I had once thought to grow accustomed to the gore. But the human suffering elicited a visceral response in me, and it did not go away.

Every Monday morning in Patient-Doctor, my group of forty students found a sheet of paper listing patients according to diagnosis on a wooden door buried in a corner of the oldest section of the hospital. Every week I selected a patient from the list and signed my initials in the "student" space to claim my patient. My classmates and I left individually to find our patients, examine them, and write up their histories and physicals. We reconvened later in the afternoon in our small groups to discuss and review our patients.

One Monday I had to choose between two patients, and I noticed that the small bowel obstruction was in the room with a patient I had interviewed the previous week. So I took the pancreatic cancer patient in the next room instead. Still savoring the last moments of my weekend, I entered the room expecting to find an elderly, sickly man. But Mrs. Mitchell was a young, blond fifty-year-old woman. Diagnosed with inoperable pancreatic cancer in January, she had a particularly dismal prognosis.

Taking her history, I learned that Mrs. Mitchell had two children. She worked in a shop in Connecticut. Except for a large incision spanning her abdomen from her navel to her sternum, she was perfectly normal on physical exam. She hoped to go home on Wednesday, and I asked her if she had any plans for when she returned home. She looked at me, and she cried. "I'm just going home to wait and see how long I have."

As I listened to Mrs. Mitchell, I thought back to David, to Dan and Steve, to Sara, to Tracy, to Natalie. This was the real pain. I didn't pass out that morning during my first surgery. But somehow it didn't seem as important anymore.

Trauma

Everyone assembled in trauma bay 2. Staff members donned white gowns over their blue scrubs and fastened them with a tie at the waist. They shielded their faces behind what looked like welders' masks. They checked the ventilator. They prepared the defibrillator. They waited.

I was in the emergency department as part of my Patient-Doctor rotations. Unlike the usual morning sessions, this one was scheduled for the evening to coincide with a busier time of day for the ED.

I watched the double doors to the ED expectantly, not sure what would come through them. "Traumatic Arrest," the page had said. "That means that someone has had a cardiac arrest secondary to trauma," explained the physician I was following for the session. "It's quite likely that we will have to open the chest to repair some damage to get his heart working again."

When the doors finally opened, two emergency medical technicians who work on ambulances (EMTs) pushed in a gurney bearing an obese man. The mountain of his stomach hid his head from view. There was a striking lack of blood for what I expected from "trauma."

"John Doe is a fiftyish man in asystole. He's been down approximately fifteen minutes. He drove off the road and suffered trauma to the head," said the EMT. "Asystole" meant that the man did not have a spontaneous heartbeat. He had been unconscious for fifteen minutes.

"Do we know who he is?" someone asked.

"We searched the car, but there was no identification. We're trying to figure it out now," the EMT said.

The gurney pulled into the trauma bay, and the waiting medical staff swarmed around the patient. Only his feet were visible at the end of the gurney, sticking out beyond the activity. A digital clock on the wall above the gurney counted off the time.

"Okay, I've got him on our monitor," said someone as the man was transferred from the ambulance's monitoring system to the hospital's EKG machine. All heads simultaneously turned to look to the right-hand corner of the bay, where the man's heart rate was on display.

Someone cut off his brown polyester pants and then the white pinstripe boxers underneath. I don't remember seeing his shirt removed, but it was gone too. In the flurry of activity, his worn-out New Balance sneakers seemed to fall of their own accord onto the ground. They lay on the floor for only a brief moment before someone wearing blue scrubs rushed in from the side and took them away. Suddenly the man was naked. His untrimmed toenails extended beyond the tips of his toes. His white, bruised thighs were uncannily still in contrast with the action around him. His mound of stomach expanded and contracted dutifully in response to the black bag placed over his mouth, rhythmically squeezing air into his chest.

In the organized chaos of the trauma bay, the medical staff tried to rally his failed heart.

"Let's give him high-dose epinephrine."

"Another three mg of atropine." Two third-year medical students I recognized alternated sets of chest compressions as their arms became fatigued from the effort.

Suddenly everybody stopped. Cardiopulmonary resuscitation (CPR) stopped. Ventilation stopped. Everyone stepped back.

"Does he have a pulse?"

Someone checked. "No."

"Resume CPR!" The flurry of activity resumed. Two people worked to get intravenous (IV) access to his femoral arteries in his groin. Someone else used scissors to make an incision in his right foot to access the vein.

"It's too difficult to get IVs in his arms. Since he has no circulation, his veins are collapsed. We're just trying to get access any way we can," the physician said.

"He's in v fib! Let's shock him," I heard someone say. His heart had revived somewhat, but it beat erratically in a rhythm that would cause death if not reversed. The defibrillator paddles came out. The electric shock from the defibrillator was the only chance of restoring a normal heart rhythm. Everyone took a step back. The scissors were left hanging, suspended by the tunnel of skin they had created.

"Everybody ready?" yelled the man with the paddles.

"Two hundred joules. Clear!" The man on the gurney jerked and then returned to his stillness. It was just like on *ER*. All heads turned toward the EKG monitor.

"Okay. It went flat. Now we have a reason to continue. Resume CPR!" said someone. His heart, still not beating regularly, was no longer in ventricular fibrillation. The trauma team redoubled its effort to revive this man.

A woman came back to the trauma area. "We found his name, but we can't locate any family," she said.

"What? The family is here in the waiting room?" someone called.

"No, he doesn't seem to have a family," she said.

The man's feet still stuck out beyond the boundary of activity, awkwardly still amid the frantic energy of the room they occupied. They defied the efforts of everyone to save them. I watched as the slightly dusky soles gradually deepened into a more profoundly steel blue, as subtly and dramatically as the ocean at sunset. I watched as death steadfastly firmed its grip on this pair of feet.

Despite the CPR, the atropine, the epinephrine, the defibrillation, the man on the gurney still didn't have a pulse.

"Okay," someone finally said. "Let's call it. Seven fifty-eight."

People ripped off the white gowns and threw their latex gloves into the trash. They collected stray tubes and wires. The body of the man lay naked and exposed in the center of the room, no longer the center of attention. Blood and Betadine were smeared over his groin and his feet where IV access had been attempted. I left to observe other events in the ED.

A little later in the evening we needed an ultrasound machine to evaluate a patient. I followed the physician back to trauma bay 2 to find one.

The man from the traumatic arrest still lay on his gurney in the middle of the room. Now he was sheathed in a white shroud that covered the gurney and draped to the floor. I was amazed by how he had shrunk all alone in the bay. His previously mountainous stomach seemed to have shriveled. Whereas before his overgrown limbs transgressed the boundaries of the trauma bay, now the vast expanse of empty room dwarfed his body. His previously incongruous feet now melted into the rest of his body. My preceptor noticed me looking at the dead body.

"Yueesh!" he said before he looked away. We quickly grabbed the ultrasound machine and wheeled it out of the trauma bay.

Loss of Language

Mr. Szanto sat in his wheelchair next to the window. A growth of gray beard covered his sunken cheeks, and a clear plastic mask, connected to a long plastic tube leading to a small machine, covered his nose and mouth. Steam hissed as it seeped out of the gaps behind the mask and swirled around his face. He watched me with his bright blue eyes as I approached him.

"It's just a nebulizer," the nurse said to me. "You can talk to him while he's getting his medication. It shouldn't take too long." Then the nurse left.

"Hi, Mr. Szanto," I said. "Did they tell you I would be coming? I would like to talk to you about why you are here and then do a physical exam. Would that be okay?" A muffled voice spoke from behind the mask, and he shrugged his shoulders. With the hiss of the nebulizer, I couldn't quite make out what he said. I asked again. "Do you mind if I wait here until you finish your medication?" He said something again from under his mask, but then he shook his head no.

I felt awkward as I sat on his bed waiting for the nebulizer to finish. The mist from the nebulizer carried medicine to open the airways deep into Mr. Szanto's lungs. He was a patient in a rehabilitation hospital that cared for patients who no longer required the intensity of hospital services but were not healthy enough to return home. I listened as the patient in the other bed rattled on and on about his illness and medical history to my classmate. Every so often rolling laughter drifted through the curtain over to our side of the room. I looked back at my patient; he was intent on the TV. I noticed his right hand clenched in a stiff fist. His right arm rested rigidly on a pillow on his lap. Then I understood.

"Mr. Szanto, have you had a stroke?"

He turned to look at me. His blue eyes sparked with recognition as he nodded.

"Are you having difficulty speaking?" Again he nodded.

A stroke in his left cerebral hemisphere had damaged at least one of his two language centers. Mr. Szanto had an aphasia. Normally the brain segregates language into two separate functions. One language center organizes the neural signal to the lips and tongue to control speech. The other center controls the cognitive side of language, allowing us to understand and create language. Mr. Szanto had clearly lost ability to form speech, but I was unsure whether his ability to understand language had also been affected.

"Were you able to speak before this stroke?" He nodded. His eyes sparkled with this communication, with this understanding.

"How old are you?" I asked. "Sixty?"

He shook his head.

"Older?"

He nodded.

"Seventy?"

He shook his head. He removed the steaming mask with his

good hand and set it on the bed. A guttural sound emerged from his mouth as he struggled to form the words to tell me his age. I couldn't understand. Again he struggled to form the words, slowing down the syllables in an attempt to convey their meaning. Still the joined sounds failed to form a word. Finally he drew his age in the air with his left hand. He was sixty-six.

Another nurse came into the room. "I think you need a shave!" she said. "I'll get you an appointment right now, and you'll look all spiffy."

He smiled his lopsided grin, and he shrugged his shoulders. He pointed toward the door and tried to speak, but his tongue betrayed him. Now he pointed more clearly toward me.

"You have a visitor," the nurse said. "Don't worry, I'll come back later." He smiled his lopsided grin, his message successfully conveyed.

Mr. Szanto was isolated by his mind. He was able to understand language and formulate ideas, but his brain prevented him from sharing his thoughts, hopes, fears, and needs. His language was limited to the words we could supply for him.

He was one of the lucky ones.

The neurology resident brought us to meet Mr. Johnson to demonstrate his particular language disturbance. Like Mr. Szanto, Mr. Johnson had suffered a stroke to the left side of his brain that affected his language centers.

Mr. Johnson looked to be in his mid-seventies. He lay flat in his bed, staring up at the ceiling. His hands rested on his chest on top of his covers. He wore a wedding band.

"Hello, Mr. Johnson," the neurologist said.

"Hello," Mr. Johnson replied.

"How are you today?"

"Good," Mr. Johnson replied.

"Okay, Mr. Johnson. I'm going to ask you a few questions. Do you know where you are?" Mr. Johnson looked back at him blankly.

"Where are you? What is this place?" the neurologist asked again.

Mr. Johnson said nothing.

"Am I a doctor?"

Mr. Johnson nodded.

"Okay. Very good. Now, can you point to the door, Mr. Johnson?"

Mr. Johnson just looked at us.

"The door, Mr. Johnson, the door. Can you point to the door?"

He raised his left arm and gestured toward the ceiling.

"No, the *door*. Can you show us the *door*?"

We looked expectantly at him. He gestured toward the ceiling again.

"Okay. That's okay. Let's try something else." The neurologist pulled his tie from behind his white coat, a reflex hammer from his pocket, and a pen from another coat pocket. "Which one is the writing utensil, Mr. Johnson? Can you point to the writing utensil?"

Mr. Johnson looked at us.

"The writing utensil, Mr. Johnson. Which one do you write with?"

Mr. Johnson pointed to the tie.

"No, which one do you *write* with?"

This time he pointed to the hammer.

"Well, I think we are finished here," the neurologist said. My classmate and I trailed the resident out of the room.

At forty-nine, Mr. James was the youngest person on the stroke service. A month previously he had been on a skiing vacation with his wife and children. His physicians suspected that a fall on the slopes caused neck trauma that precipitated a tear in the lining of his carotid artery, providing an alternate route for the blood and disrupting flow to the left side of his brain.

When we entered Mr. James's room, he lay in bed, watching

TV. He turned off the television so we could visit with him.

"Hello, Mr. James," the neurologist said. "How are you doing today?"

"Well, I feel pretty good, actually. It's been a good day."

The neurologist introduced us as medical students and explained that we were learning about the consequences of strokes.

"I'd like to be of help," he responded.

"Okay. Mr. James, I'd like you to repeat after me: 'No ifs, ands, or buts about it.' "

"No ifs, ands, or buts about it," Mr. James repeated.

"Do you know where you are?" the neurologist asked.

Without hesitation, Mr. James answered, "The hospital."

"Mr. James, could you tell us what happened to you? Why did you come here?"

Mr. James began to speak eagerly. "Well, the night the hockey game in the evening the night before. The puck my head the night the day," he said. I struggled to find the meaning in his story. He spoke fluently with appropriate eye contact and hand gestures. He looked so natural that it took me a minute to realize his story was just a random group of words strung together, and sometimes they weren't even real words.

"Did you notice anything about your speech, Mr. James?" the neurologist asked. "Because it was hard for us to understand what you were trying to say. Some of your words didn't make very much sense to us. Did you notice any mistakes?"

Mr. James looked at us, surprised. "No." He shook his head.

When we left the room, I asked the neurologist whether Mr. James understood that he had a language disturbance. "Of course not. He said he didn't hear anything wrong."

"But I mean globally. Does he understand that while he may not hear it, his speech is unintelligible?"

"Oh," he said, "I really don't know."

These people with the language disturbances. They were trapped by their bodies, thwarted by their minds. At best they were limited to the words and ideas we provided them. It felt very different from meeting patients who did not speak English. While language failed us, it did not fail them. They could communicate, just not with me. But I ached for these stroke patients and their families. In one moment, with or without warning, a transient failure of blood flow to a vulnerable region of the brain caused irrevocable damage. Each brain, damaged slightly differently, manifested its unique set of deficits. While the bodies of these people were still with us, in some ways they were lost forever.

On My Way

Now nearly finished with my second year, I suddenly recognized my transition to the medical mentality. Before, I consciously pushed myself to think like a doctor, but now I seemed to look at everyone as a patient. While waiting in line at the grocery store, instead of reading the magazines in the checkout line the way I used to, I studied the tremor of the woman in front of me. I diagnosed the homeless man with a right arm that repeatedly shot out and traced a crazy arc in the air with the movement disorder hemiballismus. I saw a man on the subway with a blond tuft of hair in the middle of his forehead and speculated that he might have Waardenburg's syndrome, a congenital anomaly causing deafness and characterized by this hair discoloration. All the illnesses I studied in school suddenly seemed to materialize in the people around me, who I had always before assumed were healthy. I didn't notice when this started happen-

ing, and I wasn't sure I liked the habit. I felt as though I had
invaded their privacy. They didn't choose to share their illnesses
with me. Yet I now looked through medical eyes, no longer
clouded by the oblivion of the layperson.

By the end of the year I began to notice this transition in my
patient interviews as well. Genny was a sixty-eight-year-old
grandmother. Her gray, curly hair was neatly in place, and she
smiled and joked with me during the interview. Genny was
having difficulty breathing. She felt faint and short of breath
while walking and intense pressure in her chest relieved by resting. She could no longer walk up the stairs to her second-floor
bedroom. She had already had one heart valve replaced just a few
years earlier. On the basis of her history, I asked her if she had
suffered rheumatic heart fever as a child.

"You know," she said, "everyone's been asking me that!"

Finally, after two years of bumbling through histories and asking patients to spell the names of their diseases, I knew I was on
the right track. After all, everyone else had asked her the same
question too.

As I studied for the Boards and attempted to cram every last
clinical detail into my head, the knowledge I had accumulated
over the last years finally began to make sense. I thrilled with
accomplishment and relief as my patients complained more and
more often that I bothered them with the same old questions.
The physical exam was becoming a little more routine and a little
less nerve-racking, although I had yet to do one without forgetting
something important. I began to think that maybe I would be a
good doctor someday.

My classmates experienced a similar surge in clinical competence. At the end of March, Andrea followed a team of interns
and residents as they examined an AIDS patient receiving the
antibiotic Bactrim to prevent *Pneumocystis carinii* pneumonia
(PCP), which affects severely immunocompromised people.

The patient had been admitted for a fever of unknown origin.
He also had a mild rash. The residents and interns generated a
long list of possible diagnoses. They busily ordered tests and labs

and more tests to help narrow down the possibilities. After they had finished their examination, Andrea asked the patient when the symptoms had begun. He told her they had started approximately one week after he began the Bactrim. Afterward Andrea pulled aside one of the residents and asked if the patient's symptoms could be caused by an allergic reaction to the Bactrim, a drug people were commonly allergic to.

"You know," the resident said, "I hadn't thought of that. But that is an excellent idea. It would definitely explain all his symptoms."

Later that afternoon the resident found Andrea studying in the library. After the test results had come back, he told her, they determined the patient was indeed having a drug reaction. She was right. They switched the patient to pentamidine for pneumonia prophylaxis to avoid the allergic response. Becoming a good physician felt within reach.

Yet as I began to feel more comfortable with my clinical abilities, I also became acutely aware of how much more I needed to learn. At the end of my first year I was most impressed by how little I knew. At the end of my second year I was amazed by both how much I had learned and how much more I needed to know.

I had heard about the dreaded "pimping" since the beginning of first year. During third-year clinical clerkships, preceptors would grill my classmates and me on our understanding of clinical issues in front of other medical students, residents, and patients. This system of pimping was intended to probe our knowledge, to demonstrate the important points, and to check that we assimilated new information appropriately. It presented us with the opportunity to flaunt what we knew or to humiliate ourselves through our ignorance.

In Patient-Doctor II, I got a small sampling of pimping. My preceptors often quizzed me on my patients' symptoms, abnormalities on physical exam, and treatments. "What are the five signs of inflammation?"; "How do the symptoms of small bowel and large bowel obstruction differ?"; "What drug do you use to treat acute gout?"

Sometimes I could answer, but many times I could not. The course coordinators encouraged us not to be afraid to say, "I don't know." But I had no pride. I said, "I don't know," with ease, even if I was clearly expected to know the answer.

That year my classmates and I received special dispensation. No one expected us as second-year students to know anything. As I walked with my preceptor for the morning to see the first patient, he or she often started to ask a question. "What are the causes of— Wait, what year did you say you were? Second? Okay then, the major risk factors for MI [myocardial infarction] are hypercholesterolemia, smoking, hypertension, family history, and diabetes. . . ."

I relaxed, relieved my ignorance wouldn't be on display at least for this morning. Then I realized that effectively the difference between second and third year was only a matter of three weeks. I would take the Boards in three weeks, and then after two and a half weeks of vacation, third-year rotations would begin on the first of July. That meant I had three weeks to assimilate all the information I was not required to know as a second year but would be expected to have learned as a third year. Grateful for the temporary relief now, I trembled in anticipation of pimping during my clinical rotations.

In the last months of second year the reality of third year slowly settled in. Before my classmates and I left for spring break, we had to turn in our rotation requests. The third year was broken into four quarters, with two taken up by the surgery and adult internal medicine rotations. A third quarter was spent on obstetrics, gynecology, and pediatrics. The fourth was "flex time," for which we chose one-month elective rotations or vacation. We listed our preferences for rotations and hospital sites, but our schedules were determined by computer. Each of the basic rotations ran simultaneously at different hospitals, so we had a wide variety of options available.

For the first time I heard about call schedules, which determined how many overnights I would spend in the hospital on call. Q3 and q4 suddenly took on very important meaning: Being on call overnight in the hospital every third night (q3) rather than every fourth night (q4) added two extra call nights to a monthlong rotation. You should never do the rotation in the area you hope to specialize in first, a fourth-year student advised, because you don't know enough yet to impress them. But you also shouldn't do it last because it's too close to residency applications for recommendations. There seemed to be tremendous urgency to select the absolutely perfect combination of rotations and sites.

Third year was shrouded in mystery. My classmates and I had heard horror stories about that year. It was reputed to be the hardest one of medical school, and the second hardest in all our years of training, superseded only by the first year of residency. I heard about the stress and the fatigue. Yet I had no sense of what the next year would be like. I didn't know what my responsibilities would be or what would be expected of me. Since the third years spent almost all their time in the hospitals, I hardly saw them. There had been no one to hand down anecdotes and experiences to give me a preview of what was to come.

Our administrators made the mistake of passing out the rotation schedules between two lectures. Rumors had circulated all morning that we would receive our schedules. During the five-minute break between the two morning lectures, some of my classmates ran out to check their mail, and the word was out. Our schedules had arrived. As soon as the news trickled into the amphitheater, the rest of us hurried out to see what fate the registrar's computer had determined for us.

I didn't hear a word of the second lecture, and I don't think many of my classmates did either. An audible hum of whispers threatened to overwhelm the speaker. Papers shuffled as we passed our schedules around to compare. I was lucky. I had gotten everything I wanted both at the site and in the time slot I chose. Carlos also did well with his schedule. While we didn't have any rotations together, our harder and easier months were coordinated.

Finally, my next year was concrete. Knowing that I would start with the surgery rotation and which hospitals I would rotate through for the next year alleviated some of the anxious trepidation, but it also made third year feel real and immediate. No longer "next year" in the abstract, now it was next month. I was eager to finish with academics and Boards and to try my hand at clinical medicine. At the same time I was terrified to leave behind the world I knew. I knew I could succeed in the classroom, but I had no idea how I would fare in the years to come.

My classmates and I took the last final exam of our regular courses two weeks before the Boards. We all spent those final weeks after our exams and before the Boards cramming like crazy. Some of my classmates put in eighteen-hour days for that entire period. Carlos and I studied together, and neither of us had the energy to put in grueling days. We had planned to study slowly and steadily so we would not need to. But I had even less patience than Carlos, and I desperately anticipated my daily aerobics class as a legitimate hourlong reprieve from studying.

When we signed up for the Boards in March, the testing company gave us a sample test along with our registration packet. Carlos and I had saved ours until the week before the exam. I planned to use mine to identify weaknesses to focus on during my last week of studying. We took the exam on his porch on a sunny morning exactly one week before the exam. When I graded my exam, I found that I had scored exactly the same as on the practice exam the school had offered four months before. After killing myself for four months, I found out that the studying had made no difference. I sobbed. I wouldn't be a good doctor after all.

After I finished crying, I was able to see the benefit of this exam score. Maybe I wouldn't do well, but I wouldn't fail either. Certainly one extra week of studying wouldn't make a difference when four months hadn't budged my score. So I didn't take my last few days of studying terribly seriously.

The exam itself was torture. Each of the two days was broken into two sessions of three hours each. The multiple-choice questions were difficult, and there were 180 of them per session, giving us only one minute per question. After a while I stopped caring which answer I chose; they all seemed equally good. I had never walked out of an exam feeling so bad despite my intense preparation. Many of my classmates felt the same way, and as we discussed the questions, we all realized how many we had answered incorrectly. Disappointed by my performance, I was nonetheless relieved to have the exam behind me.

The day after the Boards Carlos and I left Boston for a ten-day trip to Greece. We traveled to three of the islands, where we spent our days on the beach and our evenings over long dinners. While in Greece, I learned that I had gotten a contract to write this book, and Carlos and I shared a wonderful celebratory dinner looking out on a moonlit bay. We returned to Boston tanned, relaxed, and, if not quite ready to start again, at least curious about what lay ahead.

The Clinical Years

Surgery

I didn't know who she was, this, my first patient in the operating room. She was the obese black woman lying on the gurney in the first curtained cubby in the holding area. I watched as her husband kissed her good-bye before the nurses wheeled her into the operating room. Once in the operating room, she asked me to keep safe for her a small card inscribed with a calligraphed prayer. I had no idea what to do with it, but one of the nurses rescued me and attached it to her chart. The nurses transferred her to the narrow operating table and then cleaned her abdomen with a sterile brown iodine solution that dried to golden yellow. They draped sterile green towels to create a surgical field, an area shielded from any germs still lurking on other parts of her body. The anesthesiologist came with his medicines, and she drifted into a deep chemical sleep.

The operating room looked much as I had expected. It had a

white floor and green tile walls. A cream circular lamp cast a warm white glow on the woman's belly. Computer screens displaying various colored numbers and graphs monitored the woman as she slept. The ventilator hissed as it rhythmically breathed for her. Metal counters flanked the sides of the room, and a simple clock hung over the door to mark time.

"She has advanced metastatic breast cancer," the intern quickly explained to me before the surgeon began his work. "She's in for a bowel obstruction. The tumor has invaded her colon."

I watched as the surgeon made the opening incision. As the scalpel passed down her abdomen, the dark skin leaped away from the blade, exposing the delicate golden yellow layer of fat below. The fat, soft and pulpy like the inside of an orange section after the translucent skin has been peeled away, sprouted miniature pools of crimson blood. The surgeon reached in with an electro-cautery device, and the high-pitched buzz of the instrument was overpowered by the crackle of burning flesh as the crimson pools charred bubbly black. The surgeon continued to cut with his scalpel through the layers of golden fat, cauterizing the tiny vessels as they released their small pools of blood. Finally, the surgeon's scalpel exposed the bowel, concealed under all the layers of fat. As they pulled away the abdominal wall to expose the loops of invaded bowel better, I caught sight of her bubble gum pink intestine, glistening as it quietly snaked its way through her belly. The pink intestine wound its way through her abdomen hand in hand with a narrow band of globular tangerine-colored fat.

I was touched by the surprisingly thin layer of dark brown skin that yielded to the delicate gold of her fat, which in turn gave way to reveal the flaming bowel and its pale orange fat. These vivid colors and textures were so unexpected after the dusky pink-gray monochromatic organs of our cadavers that blended seamlessly one into the other.

Surgery was an intimidating rotation to begin my clinical experience with in July, and no amount of orientation was able to prepare me for my first experience in the operating room. I didn't know where to stand or what to do. I didn't know what to expect. I was thrust almost without introduction into the intense world of hospital medicine.

Surgery had by far the worst reputation for early-morning hours and long days, and there were horror stories about how mean surgeons could be. Nevertheless I had requested surgery first. I was fairly certain that I did not want to be a surgeon, so if I was going to make a fool of myself while trying to adjust to hospital life, this was the time. Plus, if I had to get up at 4:00 A.M. every day, it might as well be warm out.

The night before our rotations started, Carlos and I wondered why we had chosen this lifestyle. Carlos was also starting his surgery rotation, but at a different hospital. Tales of the grueling regimen of the program at Carlos's hospital had trickled down to us.

"You're not going to see him for the next three months. He's going to disappear off the face of the earth," one of the fourth-year students warned me.

Even though my site was reputed to be less intense, there was no such thing as a light surgery clerkship. After spending every day of the last eight weeks together studying for the boards and then vacationing afterward, Carlos and I worried that long hours in the hospital and disparate call schedules would prevent us from seeing each other.

At eight o'clock on Monday morning I walked to the hospital. It was a warm, sunny day, and I was nervous. I had no idea what to expect. I had prepared a small bag with toothbrush, toothpaste, and extra underwear in case I would not be home that night. When I walked into the main lobby, I found a group of my classmates waiting until it was time to go to the introductory meeting. Roy was supposed to be in the same rotation, but he hadn't arrived yet. He was usually a little late.

The main lobby of the hospital was small, with two rows of maroon chairs facing the wall of windows looking out onto the main entrance. The chairs were grouped in twos, threes, and fours separated by small wooden end tables, resembling an airport waiting area more than a hospital lounge. The hospital was very old, and it had expanded over the years by a haphazard assortment of added buildings. They formed a complex maze. Even after three months I couldn't reliably find my way from one building to another.

The surgery rotation was three months long and divided into two months of inpatient service taking care of hospitalized patients and one month of outpatient care. During my inpatient requirement I was assigned to the trauma team, which, in addition to its general surgical responsibilities, handled all accident victims. I also rotated through urology, anesthesiology, and the emergency room. The outpatient month was light, and I spent my time following orthopedists and ear, nose, and throat specialists through their daily schedules. To round out my hospital experience, I spent time as well in a general surgery clinic following outpatients with surgical illnesses.

General surgeons saw patients with bowel disease, ranging from obstruction and tumors to appendicitis. They also cared for patients with liver and gallbladder disease, those with breast cancer, and those who needed basic thyroid surgeries or their spleens removed.

When I arrived on July 1, I had no idea how patient care was organized in the hospital. But I quickly learned my way around the hierarchy. Inpatient care was delegated to different medical teams, divided by specialty. Each team consisted of between three and six residents and was responsible for the care of fifteen to twenty-five patients. A senior resident supervised each team, keeping track of all the patients and their care plans and helping more junior team members solve clinical dilemmas. Interns, in their first year of residency, were the workhorses of the team, scheduling patients for tests, performing basic procedures, admitting new patients, and discharging those ready to go home. Medical students worked under the interns and aspired to carry out a similar role for fewer patients. One or more staff physicians, who had already completed their training, supervised the entire team and ensured that we did not make mistakes or miss potential problems. They were also responsible for teaching. This general structure served as a system of checks and balances to teach inexperienced clinicians by giving them responsibility while providing ample patient safety. This team structure was preserved in every specialty to train residents and to care for patients.

During my surgery rotation, I was required to "pick up" patients. This meant I scrubbed in on their surgeries—I washed up, wore a sterile gown, and stood near the surgical field as I observed the procedures. I also helped write postoperative orders for medications and IV fluids, examined the patients daily, monitored their recovery, recorded any test results, and reported on their progress to the team every morning on rounds. I was expected to understand the medical issues, the operative procedures, and any postoperative complications of all my patients.

For students and interns, the typical surgery day began between 5:00 and 5:30 A.M. when we prerounded. During prerounds the interns and I gathered information about our patients and acquainted ourselves with their nighttime courses. We awakened and examined our patients, wrote brief notes on their conditions, and wrote the daily medicine and IV fluid orders before work rounds. Work rounds began at 6:00 A.M., when the senior and junior residents joined us. During work rounds the senior and junior residents, the interns, and I traveled in a group from room to room, visiting each patient on the service. The interns and I reported on the pertinent statistics and issues for each patient that we had gathered during prerounds to the senior resident, who made note of their progress and offered advice to optimize their care. After work rounds, which could last more than an hour, the residents, the interns, and I had a brief attending rounds, in which the residents updated the surgeons on the progress of each of their patients.

Finally, the interns, residents, surgeons, and I were ready for the operating room (OR). The first case began at 8:00 A.M., but we needed to arrive earlier to meet the patient and prepare him or her for the procedure. During surgeries one resident acted as first assist on the case and was responsible for helping the surgeon operate. Medical students helped retract tissue to optimize the surgeon's view and clipped the ends of sutures after the knot had been tied. I found both tasks tedious. Usually a little too short to see the surgical field well, I was too shy to bother the busy nurses for an extra stool. I usually daydreamed, and I often heard, "Cut! *Cut!*" before I realized someone was talking to me. Occasionally a student helped suture when closing the skin wound, but otherwise we mainly observed and answered, usually incorrectly, any questions that the surgeons or residents asked us.

The surgeons, the residents, and I were generally in the OR until late in the afternoon. After the last case the residents and students reconvened for brief sign-out rounds. The interns and students summarized the major events of the day for each of the patients, and the on call resident took note of ongoing issues and potential problems that might arise over the next twelve hours. After sign-out rounds the rest of the staff took care of any remaining issues with their patients and then left for the evening. The on call resident stayed to manage any patient issues that came up over the night.

I took call every third or fourth night, although the interns and residents were on call more often, every second or third night. On any given call night I got anywhere from four hours of sleep on a good night to less than one hour on a busy night. I never thought I would be able to tolerate the sleep deprivation, but it wasn't as bad as I had feared. Adrenaline and caffeine carried me easily through the day, and it was actually the next day that was more difficult. Unlike the interns, once asleep, I no longer had any obligations. I didn't have to worry about being paged; generally students weren't important enough for anyone to make the effort. But the interns had to worry about all the things that could go wrong and were woken quite often.

"You are no longer the focus," Dean Federman had told us in our orientation to rotations and our clinical education. "The patient is." He wrote this on the board, underlining "patient" twice, to emphasize his point. This was not news to me. Patients were in the hospital for treatment, not as a learning vehicle for me. No longer the tutorial cases developed for the sole purpose of allowing me to ponder and dissect in order to deepen my understanding of pathological principles, these were real people coming to us for real care. I was upset that a clinical preceptor would even think I might consider my learning agenda the priority.

Yet on the wards I soon realized that while the patient took precedence for the team of clinical care providers, my learning objectives were still all about me. I was told to be aggressive in pursuing my education. I should make sure to see what I needed to see, ask as many questions as I wanted, perform the procedures I needed to master. And, my intern told me, I should never, ever, *never* turn down the opportunity to perform a procedure if offered the chance—no matter how nervous or how unprepared I felt.

Stephen was one of the first patients I met that year in the emergency department (the first segment of my surgery rotation). He was pale and gaunt, and his blond crew cut and light blue eyes made him look younger than his twenty-seven years. He recounted a history of progressive urinary frequency that had become significantly worse over the last month, insatiable thirst, and weight loss despite a good appetite.

Although unsure how late in life juvenile-onset diabetes (now called Type I Diabetes) could develop, I was fairly certain this was Stephen's problem. The nurse pricked his finger, squeezed a drop of blood out onto the Chemstrip, and turned on the glucometer. The machine beeped to signal the beginning of analysis and counted down the thirty seconds until it announced the blood sugar level: 29...28...27...

Let it be high, I thought. I wanted it to be diabetes. I wanted to be right. Then I realized what I was wishing for. I wanted Stephen to have diabetes so I could have the satisfaction of getting the right diagnosis. I felt guilty and uncomfortable.

The glucometer counted down 5...4...3...2...1...and beeped three times to signal that it was finished computing. The screen read "571," an extremely elevated sugar level in comparison to the normal range of 70–130. Stephen had diabetes. I was right.

I struggled to find a tenable balance between patient-centered care and student-centered education now that I worked on the wards. While I provided care first and foremost for Stephen, I was also there to learn to recognize diabetes in my future patients.

Initially I was intimidated by the hierarchy, the pace of the day, and the responsibility. In spite of the minirotations I did in Patient-Doctor II through the various specialties, including surgery, the hospital seemed completely unfamiliar. My white coat, now a little less white after two years of use and wrinkled rather than freshly creased, felt new and uncomfortable all over again. I had discarded my black camera bag because it identified me as a second-year student who didn't yet belong to the wards. In preparation for third year I had stuffed my pockets with the essential equipment and books of reference material. My coat weighed heavily on my shoulders. The amount of stuff crammed into the pockets was inversely proportional to the wearer's level of training. More senior residents and physicians were easily recognized by their light and airy coats. As more information was transferred from the reference books into the brain, the manuals crowding coat pockets could be discarded one by one.

In addition to my heavy coat, I wore scrubs underneath, a new piece of the medical uniform for me. The reversible blue or green poly-cotton tops and drawstring pants were strictly utilitarian in design and indicated to bystanders the hard, messy work the wearer performed. Even the size small unisex shirt bloused uncomfortably under my coat, and I had to roll up the pants several times to keep from tripping. Nonetheless I was proud to wear scrubs, craving every concrete reassurance that I belonged in this unfamiliar world.

For the first time I developed long-term relationships with patients over days and weeks when I was intensely involved in their medical management and their emotional struggles, unlike the brief patient interactions of first and second years. As the year progressed, I played an increasing role in diagnosing patients and explaining their diseases to them, developing treatment plans, and performing simple procedures. No longer the anonymous hospital patients who consented to talk with me as a first-year student or let me practice physicals as a second-year student, these were *my* patients. I wasn't sure what to expect from these relationships and was surprised to find that the primary issues I faced during my first two years continued to dominate these new experiences. During the brief encounters of my first and second years, each patient raised a single issue. But as the depth and intensity of my patient relationships increased, the various issues merged and blended in each patient experience.

During second year I had difficulty getting used to the intimacy of touching patients. I was shy around their naked bodies. During third year I lost this sense of privacy as seeing undressed people became routine. In third year not only did I touch my patients, but I transgressed one more boundary as I caused them pain and discomfort while performing simple procedures. As a first- and second-year student I first recognized the power discrepancy of the patient-doctor relationship. But as I learned to counsel patients on treatment options, I confronted this issue personally. I had met many people who were dying during my first two years of patient-doctor, but now they were my patients and I shared responsibility for their deaths.

I accumulated a more sophisticated understanding of disease and treatment as I worked in the hospital. For the first time I could actually answer some of the medical questions from my family. But my self-doubt kept pace with my clinical accomplishments, and it was hard to recognize this growth. Most of the time I still felt impossibly far from being a real doctor.

As I began my first hospital rotation, I was relieved to find that I liked being in the hospital and working as part of a team. But as much as I enjoyed being in the hospital and caring for patients, I didn't especially like surgery. My feet still remember the dull ache of long hours of standing in the operating room. While I understood the tension, I hated the atmosphere in the OR. With some exceptions, surgeons who seemed to be calm and polite in other venues became different people when they set foot in the OR. The slightest hint of complications caused them to displace their stress onto their colleagues. After much discussion with my classmates we quickly established the pattern. The surgeon yelled at the scrub nurse, who was handing him tools and supplies; the scrub nurse yelled at the circulating nurse, who was handing her tools and supplies; and they all yelled at the medical student, who was likely doing her best to stay out of the way and retract the tissues adequately. During one long bowel surgery that I watched, a senior surgeon supervised as a resident performed the procedure. As he watched her make the skin incision, he told jokes. But every so often he became dissatisfied with her technique.

"What the fuck are you doing? You're going to screw the whole thing up! You can't do it like that!" He grabbed the scalpel from her and demonstrated the appropriate technique, but soon he slipped back into jokes as the resident continued.

The surgeons brought this aggression to their weekly M & M conferences, to discuss morbidity and mortality among their patients. These conferences provided a formal forum in which to discuss complications and deaths with the goal of improving procedures and modifying settings that might have lent themselves to error. But the surgeons were often punitive. One surgeon discussed a patient who had developed a wound infection after a hernia repair. This was a common complication, more frequent in obese patients, that could occur no matter how careful the surgeon. This particular surgeon had used mesh, an accepted procedure, to reinforce the abdominal wall and prevent the bowel from slipping through, causing the hernia to recur. "I never use mesh," the surgeon running the conference said. "That's why I never get wound infections."

He thought for a few seconds. "Dr. Silverstone," he asked another surgeon in the audience, "do you use mesh when you do a hernia repair?"

"No, I never use mesh," Dr. Silverstone responded.

"And, Dr. Black, do you use mesh when you do a hernia repair?"

"Never," Dr. Black said.

The surgeon presenting the case was flustered. He finished his presentation, reporting that the patient had recovered and was doing well.

I felt bad for him. I later learned he was a new surgeon, but there had to be some better way to educate him on the side effects of mesh without subjecting him to public humiliation. I attended M & M conferences weekly for three months, and most of the sessions were more pleasant and friendly. But this aggression lurking beneath the surface, to be released at any moment, frightened me.

I found the surgical issues intriguing and the patients compelling, and I liked my rotation. But at the end of the three months I was ready to move on. I knew I could not tolerate this atmosphere for a lifetime.

Nearly a month into my surgery rotation, I received my board scores. In the midst of my hectic day-to-day schedule I had almost forgotten about them. One of my classmates told me at lunch that his scores had arrived that morning, and suddenly all the anxiety I had felt on turning in my exam rushed back. It was a slow afternoon, and my intern let me run home to check my mail. Sure enough, there was an ominous manila envelope awaiting me. When I opened the envelope, I was looking at the back of my score sheet, and there was an orange graph showing my distribution on the curve in each of the various categories of questions. It looked as though my scores were at the lower end of the scale, well below passing. My hands shaking, I turned the page over and was relieved to see "PASS" at the top of a column of numbers. I turned the paper over again and realized that I had read the graph backward. I rushed back to the hospital. I was ecstatic. I had passed.

With that exam finally under my belt, I reentered the hospital with new confidence. I had proved to myself that I had mastered the first two years and was now ready to move on and become a doctor.

Procedures

On the first real day of my surgery rotation, after our day of orientation, I arrived in the emergency department exactly at 8:15 A.M. as I had been instructed, dressed self-consciously in scrubs. I walked to the central station in the acute section, where I had been told the residents would assemble before the lecture. The acute side, which dealt with more serious or immediately life-threatening issues, was organized around a central station. The patient "rooms" were actually curtained cubicles on the perimeter, and two of the cubicles were equipped with ventilators and crash carts stocked with tools to resuscitate dying patients.

When I arrived, there was no discernible group. Nurses in blue scrubs and haggard-looking residents, identifiable by their white coats, rushed to update computer files, record entries in patient charts, and discuss lab results. I stood in front of the central desk, trying to look inconspicuous in my inactivity but realized that I stuck out like a sore thumb. I anxiously awaited my classmates so we could bond in our communal awkwardness.

I didn't have to wait too long for them to arrive, and soon there was a definite nucleus of people milling around the front desk. While the residents basically ignored us, I felt much less conspicuous now that I was not the only one aimlessly waiting.

The lecture was uneventful. My classmates and I sat in the back, listening quietly, determined to fade into the background. After the short lecture we were divided among the different areas of the ED. I was assigned to the subacute side.

The subacute side of the emergency department was a little less overwhelming. Although people were sick, they generally did not suffer anything immediately life-threatening. Unlike those in the acute section, the patient rooms had real doors and looked more like the traditional examination rooms I was used to, with an examination table, a lamp for pelvic exams, a sink, and a bank of shelves filled with a myriad of medical supplies.

The senior physician in charge that day was busy caring for some early-morning patients, and the interns worked on the computer and filled in charts. The nurses ran between examination rooms, drawing blood, starting IVs, and administering medications. For the second time that morning I waited, trying not to look so out of place and desperately hoping someone would come and tell me what to do.

The senior physician came over shortly with interns in tow, Jason and Chris. "We have a lac in room ten. That's medical student bread and butter. Why don't you take it, Ellen?" he said. Someone had a deep wound that required stitches; "lac" is short for "laceration."

"This is my very first rotation," I reminded him. "I don't know how to suture."

"Oh, okay." He seemed disappointed. "Chris, why don't you take her and teach her how to suture? She can take the history." Finally, after what felt like an eternity, I was integrated into the system. Or at least I had someone to watch and something to do.

Chris took me in hand, encouraging me to sign my name in

the space labeled "doctor" on the patient list and to interview and examine the patients myself. He reviewed my first few cases to point out omissions and mistakes before I presented them to the senior physician. He went with me for the particularly challenging patients and called me in to look at interesting clinical findings on his patients. He showed me how to use the computer system, how to write a note in the chart, and quizzed me on important clinical facts and filled in the gaps in my knowledge. Just before I went home on my first day, Chris called me aside.

"Do you have any questions from today?" he asked. "I was in your place only two years ago. I still remember what it's like. So, if you have any questions that you don't feel comfortable asking the senior physician, you can feel free to ask me."

Despite all efforts to integrate me into the system, however, I required so much supervision that I was more hindrance than help. One afternoon I returned from a lecture to find the ED overflowing with patients. The examination rooms were full, stretchers lined the halls, and more people waited outside to be seen. The interns frantically ordered tests and saw patients. The senior physician bustled from room to room, sorting out who could be discharged, who should be admitted to the hospital, and who still needed more tests before a decision could be made. The nurses desperately tried to keep track of blood cultures, urine samples, and medication requests. But because every patient was already being seen, and more important, because I had few clinical skills to offer, I stood in the central area waiting for things to calm down enough for me to be involved again.

I envied the interns. I envied their role, their necessity, their knowledge. While I definitely did not desire their hours or their exhaustion, I wanted even a taste of their function in the system.

Now, as we began third year, my classmates and I placed a premium on learning procedures. This new ability distinguished us from the second-year medical students we were just three weeks before. It didn't really matter what the procedure, just wielding a needle or even a Steri-Strip bandage was enough.

I found Roy finishing lunch in the cafeteria just before our afternoon lecture. He was on the second to last day of the anesthesia portion of the three-month surgical rotation. He looked dejected, staring at the uneaten food on his plate.

"What's going on? Why are you so depressed?" I asked him.

"Oh," he said, "nothing. It's just that I haven't been able to intubate yet, and I feel like I should have already learned how." Roy wanted to learn how to insert a breathing tube into a patient's trachea so that a ventilator could breathe for the patient once the anesthetics took effect.

"I know I should've learned to intubate by now. Every morning the resident in charge asks me if I have done an intubation, and I have to say no, even though I've been taught how every morning. Once I say no, they immediately start to teach me the steps all over again. But today I got to try, and it was a difficult intubation. I couldn't do it. It doesn't bother me that I couldn't intubate that patient, but if I don't learn to do it now, when will I?"

After I compared notes with two other classmates also working in the ED, it became clear to me that I was not learning procedures at the same rate as they were. One had already drawn blood once, watched two lumbar punctures, and removed stitches. The other, who was particularly assertive about asking to do procedures, had stitched two lacerations, done one lumbar puncture, and inserted a Foley catheter into the bladder of a confused and disoriented alcoholic. I hadn't even drawn blood yet. I felt bad about my lack of initiative. Was I failing in my responsibility to educate myself?

Finally, just when I had resigned myself to never doing a procedure, I got my chance.

Vivian sat on the examining table. She was a black woman in her early forties. She had already removed her top and struggled to cover herself with a paper gown. She did not seem to be in any discomfort, although she reported a nosebleed earlier in the day and a painful left armpit.

Although Vivian was not officially my patient, the senior physician called me to see her with him and the intern. Dr. Rose was one of my favorites in the ED. He was in his mid-forties and had unkempt curly brown hair. His tie was always a little loose and often hanging at a crazy angle, and he continually pushed his wire-frame glasses back in place. Dr. Rose's skin was affected by vitiligo, a disease that destroys skin pigmentation. The vitiligo had created subtle patches of unpigmented white skin that barely contrasted with the pale ocher of the rest of his skin. As haphazard as he appeared, he was a great doctor. He was fond of pimping and a fountain of trivia. But his manner was kind, and I never felt bad if I didn't know the right answer. He spent a lot of time teaching me, reviewing my patients, and answering my questions.

Dr. Rose briefly asked Vivian about her nosebleed and then pulled out headgear from the wall behind him. He fitted the apparatus over his head and made a few rapid adjustments. A light mounted between his eyebrows illuminated Vivian's nostrils as he inserted the nasal speculum and inspected first the affected right side and then the left.

"Oh, this is really no problem, Vivian," he told her. He took off the headgear. "This is just caused by dryness. It's completely normal. You can use a saline spray to alleviate the dryness. That should take care of the problem."

Dr. Rose turned to the intern and handed him the headgear. "What do you think?"

The intern put on the headgear and quickly adjusted it. He inserted the nasal speculum, and the light fell directly into Vivian's now-flared nares. "Yeah, that's what I think too." Then he passed the headgear to me.

I slipped the headgear into place, and Dr. Rose adjusted the diameter to fit me more tightly. I looked through the eyepieces and was dismayed. I felt as though I were looking out a window between the slats of a venetian blind, except in duplicate and blurry. I couldn't figure out which knobs everyone had been turning to achieve the optimal view. Dr. Rose realized I was having difficulty.

"Here, this knob adjusts the width, this one does the focus, and with this one, you can change the angle of the mirror so the light is directly in your line of vision," he said, rapidly demonstrating the effects of each. However, since the headpiece was already on my head, I couldn't see which knobs he was turning.

I fumbled for the knobs again, desperately hoping to remedy the situation promptly and without further help, but I was not very optimistic. Finally Dr. Rose showed me where the knobs were again, this time more slowly. Then, as I adjusted the knobs, the world merged into a single image and shifted into focus. Suddenly everything was clear and magnified.

Then I tried to examine the nostril. I inserted the nasal speculum as I had seen Dr. Rose and the intern do before me.

"Is your light in the right place?" Dr. Rose asked. "Because it looks kind of high from here."

Instead of being coordinated with my line of vision, my light was centered squarely in the middle of her forehead. I readjusted the mirror, reinserted the nasal speculum, and miraculously, I was finally looking inside her right nostril. In the front portion of her nasal septum, I saw a dark splotch I was willing to convince myself was the blood clot resulting from her earlier nosebleed.

"Do you see it?" Dr. Rose asked.

"Well, I think so. In the anterior part of the septum, right?" I looked a little longer.

"Are you sure you see it? Describe what you're looking at," he said.

I told him about the dark splotch.

"It's okay if you don't see it. You shouldn't say you see something if you really don't."

"Well, I really think I see it," I told him.

"Look at the other side and tell me what you see." I looked at the other side, and I saw a completely unremarkable nasal septum, which I told him about. "Now look back at the other side again, and tell me what you see."

I looked back and again described the view. Finally I managed to convince him that I had indeed seen the blood clot. Relieved, I removed the headgear and returned it to its hook on the wall.

Next, we turned to her painful armpit. She raised her left arm to reveal a tender swelling. Dr. Rose poked at the swelling briefly and stepped back. "Well, you have an infected sweat gland there. We'll anesthetize the area and then stick it with a syringe to drain off some of the pus. Once we get some of the fluid out of there, you'll feel a lot better. Some antibiotics will take care of the infection."

Dr. Rose and the intern left the room in search of a syringe and anesthetic, and I trailed out after them.

In the hallway Dr. Rose explained the procedure to me. "We're going to use ethylene chloride as an anesthetic. It's actually not an anesthetic. It cools the skin and serves primarily as a distraction. It tells the patient that we take her pain seriously without being a real anesthetic. But you've only got one shot. After that the patient realizes that this doesn't really do anything."

Then he turned to me. "So, Ellen, you gonna do this one?"

"Sure." I tried to sound confident. I knew I shouldn't turn this opportunity down.

Dr. Rose detailed the steps of the procedure for me. "Just stick the needle directly into the middle of the abscess and pull back." He took the syringe and showed me how to attach the needle. He indicated how deep the tip should go. He demonstrated how to apply the anesthetic. "You have to be careful aiming it. Although the bottle looks like the flow is horizontal, like a Windex nozzle, it actually comes out straight," he said.

I was terrified my hands would shake.

The intern and I returned to Vivian. I had just displayed my inexperience in front of this patient. I couldn't believe she would allow me to drain her abscess, but she didn't seem concerned. The intern and I prepared her for the procedure before Dr. Rose returned. I carefully cleaned her skin with two alcohol pads and then cleaned it once more just to make sure. The intern left to find the senior physician, leaving me alone with Vivian.

"So, you're a medical student," she said, trying to make conversation. "How many years do you have left?"

"I'm a third year," I told her, trying to reassure both of us. "I've finished two years of training, and I have one more after this." I tried to make myself sound more competent than I had conveyed in the nasal encounter. I felt the familiar sick emptiness in the pit of my stomach. My hands shook as I waited for the intern and Dr. Rose to return. At that moment I was sure I was a lot worse off than Vivian.

Finally Dr. Rose returned with the intern. We positioned Vivian, and I cleaned the area with alcohol wipes one last time. Dr. Rose passed me the ethylene chloride. I aimed for the abscess but made exactly the mistake he had warned me about. Vivian winced as several drops from the misdirected flow landed in her eye. I readjusted my aim, and the flow landed squarely on the abscess this time.

Dr. Rose took the bottle from me and added another hefty dose. "Do it," he said.

I uncapped the needle and carefully inserted it into the center of the swelling. I had managed to calm myself, and my hands were steady. I pulled back on the syringe, and a few drops of yellow-green pus flew into the syringe.

"Advance further," Dr. Rose said.

As I pushed the needle in farther, Vivian winced in pain. I continued to pull back on the syringe, draining a small amount of pus.

"Okay," Dr. Rose said, "that's enough. Pull out."

Relieved, I pulled the needle out of the abscess. I put a piece of gauze to catch the small drop of blood at the surface and secured it with some tape. The intern, who stood behind Vivian, gave me a thumbs-up. Dr. Rose discussed further treatment with Vivian, and then we exited her room. As we walked down the hall back to the central desk, Dr. Rose turned to me and shook my hand.

"Strong work," he said. I had successfully completed my first procedure.

Difficult Patients

Eleanor had salt-and-pepper curly hair. She was a round, soft woman with large brown eyes and big glasses with pastel frames. On her good days she put on rosy pink lipstick, using the mirror that unfolded from her tray. White plastic rosary beads lay coiled on the corner of her tray. Every morning on rounds they rested, untouched, in the same position.

By the time I met Eleanor, the surgical team had already tired of her complaining. Frustrated by her crotchety anger, the residents were annoyed by her perceived unwillingness to get well.

One week previously Eleanor had been in a car accident. Immediately after the accident radiologists noticed a tiny contusion of her spleen, and her chest X ray raised suspicion of injury to the aorta. But further evaluation revealed no problems, and the team was prepared to let her go home until she complained of severe back and pelvic pain. She screamed in agony whenever a physician grasped her by the hips to assess her pelvic stability.

Concerned by her symptoms, the team performed all the appropriate tests. But X ray and CT scan (also known as CAT scan) showed no bone injuries and no evidence of bleeding. Still, Eleanor continued to complain of excruciating pain. Every morning on rounds she demanded a diagnosis, and every morning Pat, the chief resident, could tell her only that the tests were inconclusive and that she likely had either a tiny fracture too small to be seen or normal lingering pain after the jarring accident. Either way, the treatment would be the same, so further exploration was not deemed necessary by the medical staff; they did not want to subject her to unnecessary radiation for further bone scans or spend unnecessary health care dollars. The residents encouraged her to work with the physical therapists to improve her mobility. As the days with no concrete diagnosis accumulated, she and her husband became more and more dissatisfied with the poor quality of care they believed she was receiving. She said to me one afternoon, "If he [the chief resident] tells me one more time that I don't have a fracture, I'm going to smack him!"

Eleanor did not like Pat. He was thickset with heavy brown hair and square wire-frame glasses. He rarely smiled. Bent on efficiency, he didn't like small talk that impeded rounds, and he paid little attention to me. I liked him just fine. I had heard the horror stories about rude, imperious surgical residents, and Pat seemed a far better alternative.

On my second morning on the surgical team we arrived at Eleanor's door at the regular time for rounds. This was to be her last morning with us before discharge to a rehabilitation facility later that morning, and the team was none too sorry to see her go. We dallied outside her door, prolonging our discussion to delay the inevitable confrontation over her lack of diagnosis and our inadequate care. But we entered the room to find her writhing on her bed, gulping for air, and exhaling each breath with a grunt as she struggled to force the air out. One nurse held back Eleanor's hands as she haphazardly tried to push away the clear plastic oxygen mask covering her nose and mouth.

"Eleanor, it's oxygen. It will help you breathe!" The nurse tried to reassure her.

Eleanor's dark eyes frantically darted about the room.

Pat rushed to the bed. "Eleanor! It's Dr. O'Malley." Her eyes turned toward him. "Eleanor, what's happening? Can you tell me where you feel pain?" She looked away again without answering. "Eleanor! Do you know where you are?"

She looked toward him again. "I . . . can't . . . breathe. . . . I . . . can't . . . breathe!"

The nurse measured her blood pressure at only 70 mm Hg, much lower than the normal range of 100–140 mm Hg. Her abdomen was markedly distended. Pat ordered stat hematocrit and arterial blood gas (ABG) values.

According to the nurses, Eleanor had complained of constipation and back pain over the night before. The nurse gave her a laxative with little result. This morning, while straining on the commode to have a bowel movement, Eleanor became dizzy. The nurse helped her back to bed, and the intern ordered IV fluid hydration. Her blood pressure initially responded to the therapy but then dropped again. Her condition rapidly deteriorated. Her hematocrit of 20, the percentage of red blood cells, showed her to be very anemic, especially in comparison to a count of 27 the previous evening. In conjunction with her distended belly, this made it apparent that Eleanor was bleeding into her abdomen. She was in hemorrhagic shock.

"I . . . can't . . . breathe!" She became more and more frantic. Pat ran off to arrange transfer to the intensive care unit and to request operating room time in the event that she required an operation to search for the site of the bleeding and repair it.

"Ellen, you stay here with Dave [an intern] and help him," he instructed me.

I approached Eleanor's bed and grasped her hands to prevent her from dislodging the oxygen mask and from disrupting Dave's efforts to measure her blood pressure with the doppler machine.

As I touched her hand, Eleanor turned to look at me, and I saw terror in her eyes. She held my gaze for only a brief second, and then her eyes continued on their random, frenzied path, darting around the room. Her head was surrounded by a dingy gray halo of sweat barely distinguishable from the gray-white overwashed hospital pillowcase.

From time to time Pat reappeared. "Eleanor! Eleanor! It's Dr. O'Malley! Eleanor, do you know where you are?"

Now she no longer turned to the voice; she no longer responded to her name. She continued to writhe on the bed and struggle for breath.

Finally, transport to the intensive care unit was arranged. The nurses, Dave, and I struggled to disentangle the IV lines. We disconnected her from wall oxygen and reconnected her to a portable tank. The elevator was called, and soon we were at the ninth floor.

The anesthesiologists and intensive care unit staff were assembled and waiting for her in room 5. They immediately went to work, hooking her up to cardiac and pulmonary monitors, attaching her to wall oxygen, and placing a central line in the large veins of the shoulder to give medications and fluids rapidly. As everyone worked rapidly around her, the pungent smell of feces infused the room. No one seemed to notice as they continued to install lines and monitor her various pressures and measurements. As the effort of breathing increased, the anesthesiologists intubated her to let her rest. Eleanor was now somnolent and unarousable. But whenever anyone pressed on her upper left abdomen, her eyes opened wide, looked around, and then closed again.

A nurse acknowledged the scent of stool. "Well, at least she finally had her bowel movement!"

Somewhere along the line the decision was made to perform an emergency laparotomy. It seemed only a short time after she was finally stabilized before we left for the OR. Eleanor had to be disentangled from all her lines and disengaged from her monitors and oxygen and reconnected to portable units all over again for the trip downstairs. The transfer went smoothly, and we were soon in the OR. The nurses disconnected all the portable units again and hooked her to new monitors.

"I'm really sorry, but she was incontinent of stool upstairs, and we came down here so quickly that I didn't have time to clean her up," the ICU nurse apologized. "She's still a mess."

The surgeons quickly cut through her skin to access her belly, and I watched, standing on two stepstools in the background, as the surgeon and Pat emptied two liters of blood from her abdomen. The blood was sent through various filters and collecting tubes and a centrifuge as it was prepared for reinfusion into her veins. The surgeon and Pat pulled out handful after handful of shiny maroon clumps of clotted blood. They retrieved the shattered pieces of her ruptured spleen, the source of all the bleeding and clots. Her problem was solved.

Only a day later, deemed stable and ready to recover from her accident, Eleanor returned to her old room on the floor after a brief stay in the ICU. After holding her hand as she fought for life yesterday, I felt a strong bond with her. I had also shared the anxiety of the caregivers before we could pinpoint her problem, the concern as we struggled to stabilize her blood pressures, the worry as her mental status deteriorated and she became unarousable before our eyes. I knew that if she had ruptured her spleen after she left the hospital, she likely would have died. I felt personally engrossed in both her medical and her existential crisis. So I was eager to pick up Eleanor as one of my patients now that she had returned to our service.

I went into her room that afternoon eager to discuss her experiences from the previous day. She sat quietly in bed, her area dark and quiet except for the hum of a machine that I couldn't identify. But she was unhappy. She was having hip and back pain.

"But this is pain you've been having all week, right?" I asked her.

Oh, yes, she assured me. But she was very miserable, she complained, and the doctors weren't doing anything for her. "That damn Dr. O'Malley. So self-confident of himself. If it were him in this bed, I can assure you that he would act very differently."

"But yesterday," I prodded. "How are you feeling after yesterday? You gave us quite a scare, you know."

"Oh, that. Yes, I gather that I was very sick. I almost died. But I can't remember any of it. Except that now I have this big incision. But I've had surgery before. I know what that's like. It's this hip pain that's the problem. I can't go home like this!"

"That must have been very frightening," I persisted.

"Well, not really. I can't remember it, so it's like it never happened."

"Well, we can talk about the pain tomorrow morning on rounds. The doctors are working hard to find an answer for you, although I know the slow progress is frustrating. I just wanted to come and see how you were doing."

I knew she was worried about her hip pain, and I understood that being in the hospital and nearly dying would not predispose most people to be on their best behavior. Nonetheless I was irritated with her. Why couldn't she acknowledge how sick she had been? Why couldn't she be thankful to be alive? I understood that her confidence in us might be shaken when we had medically certified her to leave, arranged for her to go, and then she suffered this life-threatening event. But I wanted her at least to be grateful that we managed to rescue her.

But I also felt bad for feeling this way. How could I have all these expectations? What made me think that she wanted to share her experience of the day before with me? Why should I deserve that or consider it my right?

I had met frustrating and irritating patients during my first two years of medical school. I knew I wouldn't like all my patients, but I had been confident that I could break through that irritation to form a meaningful bond. Now I wasn't so sure.

We heard about the pain, as promised, the following morning on rounds. Giving in to her need for a specific diagnosis, the doctors ordered more and still more sophisticated tests. But

Eleanor expected to be able to walk and was angry that all these doctors couldn't diagnose the pain and give her medicine or surgery to take it away.

Every day Eleanor was a belligerent face awaiting us on rounds. "Good morning, Eleanor. It's Dr. O'Malley and the rest of the team come to see you. How are you doing this morning?" Pat asked.

"Well," she said, "if you say I'm doing well, then I must be doing just great." She offered a sarcastic smile and set her mouth into a grimace as she defiantly waited for Dr. O'Malley's next move.

"I want to hear how *you* think you're doing," Pat responded.

"Well, I can't walk. That's one thing. My legs, they don't work."

"But we hear from the physical therapist that you've been out of bed and getting around."

"Well, if you can call that thing I do walking. I just throw my weight from one side to the other as I scream in pain just to get a few steps to the john. So, if you can call that walking . . ."

"I understand that you are not walking the way you would like to be, but that's to be expected. You just went through a big accident and your spleen ruptured. We would expect you to have this kind of musculoskeletal pain. You just need to keep working with the physical therapist, and things should get better."

Her husband, who was sitting in the chair next to her, joined in the conversation. "Well, why haven't you done all the right tests? Why can't you tell us what the problem is? This is outrageous. She can't go home like this!"

"John, don't complain to them." His wife pacified him. "They're trauma surgeons, that's what they do. They did a great job on the spleen, and soon I'll be in a better hospital where I can see real doctors."

Whereas we used to dislike visiting Eleanor's room every morning, now we anticipated entering her room with true dread. Before, I had been less annoyed with her than had the rest of the team. But with each passing day I absorbed more and more of the team's attitude toward her. I tired of hearing about her hip pain and grew suspicious that she was malingering. Every morning Pat said at the beginning of rounds, "I pray to God they find a bed in rehab for Eleanor today."

Another morning we were a little later than usual, and breakfast had already been served. Eleanor sat in bed with a tray of eggs, oatmeal, juice, and coffee resting on the table in front of her. Pat, according to the usual routine, asked Eleanor how she was doing.

"Well, I must be doing just great since you think so," she said. She looked at him for a few seconds and then belligerently took a bite of her oatmeal. In her brusque, angry movement, a drop of oatmeal splashed onto her glasses and congealed into an opaque white tear.

"I spent all day yesterday figuring out exactly where my pain is, and I circled it with a marker. Don't you want to see it?" She proudly displayed a black circle exactly over her right hip. "So what are you going to do about that?" she asked.

"Well, Eleanor, I just don't know what more we can do for you. You've had all the appropriate tests and then some. I know you're not happy with the care you've received here. I hope in the next few days we can transfer you to another facility where you'll feel more comfortable. I think we've done just about all we can do for you."

"But what I want to know is, why haven't I been seen by a neurologist or even an orthopedist!"

"The orthopedist came two days ago. You refused the tests he wanted."

"I didn't talk to any orthopedist."

"Well, maybe you don't remember him, but he came by two days ago. You refused the bone scan. Do you want to see him again?"

"Yes! And why haven't I even had a real physical exam? How can you tell anything with these fancy tests when no one has even *examined* my legs?"

"Okay, you want an exam? Okay, right now, Eleanor. We're

going to do an exam right now." Pat ripped the blankets off her legs and put his hands on her feet. "Push down, like on a gas pedal. Now back the other way.... And again.... Okay, now lift your leg.... Bend your knees." While she had noticeable weakness, her limbs were completely functional. "Okay, your legs work."

Eleanor opened her mouth in indignation. "How can you say my legs work? Have you seen what you call walking? That is not walking, I'm sorry! How can you possibly say my legs work? If this were you in this bed instead of me, this kind of walking would be unacceptable!"

I watched the misunderstanding happen. Pat used "work" to mean functionally and neurologically intact, while Eleanor understood "work" to describe her usual ability to walk as she had before the accident. But both were so frustrated and angry that neither could recognize the discrepancy.

"Eleanor, the orthopedist will be in later to see you," Pat said halfway out the door as he led us out the room.

That afternoon I felt compelled to check in with her because I had been avoiding our daily chat for days. I was surprised to find her in a wonderful mood. She had just finished applying a fresh coat of rosy lipstick, and she was laughing with her husband.

"I *do* have bone fractures," she told me. "They found three of them on the scan. They have to do a few more X rays tomorrow to determine the severity, and then I am free to go to rehab."

This interaction shook my confidence in my ability to respond empathically to my patients. In my first two years of medical school I encountered patients that I didn't particularly like. But I always managed to respond objectively, to distance myself from the aspects I didn't like and find a common bond through the features I did like. Now, in my first long-term patient relationships, I was not able to preserve that attitude. I just plain didn't like Eleanor and avoided visiting her. How would I ever be able to extricate the personal from the clinical to be a good doctor for all the types of patients I would encounter?

The next morning Eleanor was particularly excited for rounds. She already wore lipstick at 6:30 A.M.

"Good morning, Eleanor. How are you doing today?" Dr. O'Malley asked as usual.

"Well"—she gloated—"I have three fractures! Two here"—she pointed to her pubis—"and one here, right over my hip!" She offered a rosy pink, satisfied smile.

"I'm glad the bone scan showed something and you finally got a specific diagnosis, Eleanor," Pat said. "So, rehab today? I think that should be the plan after the orthopedists get their X rays this morning," he said.

We followed him out of the room for the last time. No one looked back.

"Oh, come on. This is really cool. You should see this," a classmate said to me during my first afternoon on the team. It was her last week on the team, and my first. A surgical intensive care unit (SICU) staff physician and several nurses headed toward room 5. She motioned for me to follow. But when we reached the door, it became clear that no procedure would occur in room 5.

"It looks like they're not going to change the bandages right now. The guy in that room has a totally exposed hip. You can see all the muscles. It's great anatomy," she told me.

The intensive care unit was on the ninth floor. The nurses' station ran the length of one side of the hallway, and the patients' rooms ran the length of the other side. Unlike on the regular floors, there were no doors or walls here. The walls to the rooms were windows, and the doors were curtains, to allow constant observation of these critically ill patients. Inside the rooms, patients lay inertly on the beds, many with the rhythmic hisses of the ventilators forcing air in and out of their lungs and tubes emanating from their bodies. Few were awake or alert. Each patient was assigned his or her own nurse to administer medications, monitor for signs of danger, and track vital signs. Each room had its own TV, and although virtually none of the patients could watch, nearly all the TVs were turned on. They formed a bizarre backdrop for the events of the SICU: On Saturday mornings cartoons blared as we cared for our critically ill patients. I later learned they were intended to help orient patients to time and place.

I didn't meet Roger until the following morning on rounds. He seemed miniature in his oversize bed. A square of white sheet delineated his space in the middle of the blue air mattress. He was a sallow man with hair, eyes, and skin a uniform pasty yellow-brown. White net underpants stretched across blue sterile towels diagonally from the crest of his left hip to his groin, leaving his right hip and penis exposed. The leg holes of the net underpants revealed two ovals of worn navy blue where his left leg should have been.

Roger was an insulin-dependent diabetic with a history of IV drug use, Chris, the intern, explained to me in a whisper. By coincidence, Chris from the ED and I were assigned to the same team again. Roger had undergone an above-the-knee amputation of his left leg just over a month ago because of vascular insufficiency, a long-term complication of diabetes. He then contracted necrotizing fasciitis in the remaining stump, the dreaded flesh-eating strep skin infection. The bacteria had eaten away at the skin and denuded the entire region.

"Good morning, Roger. It's Dr. O'Malley and the rest of the team. How are you this morning?" Roger mumbled something unintelligible in response. "Did he get his morphine yet this morning?" Dr. O'Malley asked the nurse.

"I just gave him four milligrams in preparation for the dressing change," she answered.

The team members donned gloves and went to their routine corners of the bed. Not yet part of this daily ritual, I stood awkwardly in the corner. Dave, the other intern, collected gauze, two bottles of saline, and three blue-wrapped sterile towels. My classmate and the nurse turned Roger onto his right side, and Pat began to disassemble the dressing. Roger issued a steady stream of unintelligible complaints and fought to roll back to his original position.

Pat pulled back the blue towels to reveal layers and layers of gauze. He peeled off the strips of gauze to expose the stinking pink flesh below. In a matter of seconds, nothing shielded the mottled pink and red muscles from view, and the sickly sweet smell of infection pervaded the room. At the perimeter of the nude hip, an abrupt cliff separated the healthy skin and superficial fascia above from the decomposed region a half inch below. Pat reached between the gluteus maximus and gluteus medius muscles and vigorously rubbed his hand back and forth to remove any fibrous adhesions threatening to develop into scar and fuse the muscles. Roger groaned in pain. Pat lifted a thick stump of muscle of indeterminate anatomic origin to expose the pelvis hidden below. He poked his fingers deep into the crevice to palpate the bone and attempted to clean away the yellow ooze emanating from this suspected focus of osteomyelitis, infection deep within the bone. The dull white stump of Roger's femur with its congealed maroon center and ring of white bone was visible at the tip of the remaining five inches of his left leg. The stump of bone was surrounded by wedges of thick pink-red muscle. Then Dave squeezed out the excess saline from the new rolls of gauze as Pat carefully wound the soft strip through the crevices of exposed muscle. He lifted the gluteus maximus with one hand as he inserted gauze with the other. He covered the gauze with fresh blue sterile towels and secured them with a new pair of white net underpants. Roger was finally released to his former position, and the dressing change was done for that day.

Roger's hip was undoubtedly the most gruesome sight I had ever seen. I knew I was expected to watch dispassionately, without judgment or disgust. Yet I wanted to close my eyes, to plug my nose. I wanted to escape this horror that saturated my consciousness against my will. I wanted to erase this vision and prevent it from haunting me. But in my horror I was also fascinated. I wanted to see the naked fetid muscle beneath the sterile dressings. I wanted to feel behind the gluteus maximus. I wanted to see the yellow pus of osteomyelitis. I wanted to know the sickening scent of infection. I could not turn myself away.

As the week wore on, the dressing change became routine for me too. The mottled pink muscles grew less shocking, the smell less overwhelming, and his groans and complaints less wrenching. Roger's condition improved, and by the beginning of the next week he was transferred out of the SICU to the floor.

On the regular floor, no longer so heavily sedated, Roger was bright and alert. He began asking questions about his wound, what treatment he needed, and what we did to him every morning. Although he was in a double-occupancy room, he had no roommate. His blue air bed stood on the far side of the room closest to the windows. He lay on his right side in the center of the bed, the white net underpants on his left hip, and his remaining pasty, atrophied right leg stretched out toward the foot of the bed.

We often arrived at Roger's room in the morning to find both televisions blaring at full volume and the room lit to maximum brightness with the harsh white light of fluorescent lamps. One morning we found him watching step aerobics. The blond woman in the pink leotard overenthusiastically led the television audience in three-knee repeaters and around-the-worlds exactly in sync with the pink blond woman on the other screen.

We continued the daily ritual of dressing changes on the floor. Now that I was the only medical student on the team, I took over the role of turning Roger and positioning him for Pat. I stood opposite from Pat and grasped Roger on the curve of his left ribs. When he cooperated, Roger reached over for the right-sided railing with his left arm, and together we hoisted him onto his right side. Now that I stood on the opposite side of Pat and the naked hip, I no longer saw the exposed muscles as the resident unwrapped and rewrapped the dressings.

Beyond the daily dressing change, my first real interaction with Roger was an argument.

In addition to his hip problems, Roger had kidney troubles.

Because of his poor kidney function, potassium accumulated in his body. His potassium levels had been running dangerously high, putting him at risk for a potentially fatal irregular heart rhythm. One evening the nurse tried to persuade him to allow a blood draw to check his potassium levels. He vehemently refused. The nurse called Chris to resolve this issue, and I went with him.

Chris warned Roger that he could have serious heart trouble or even die if he refused this test. "That's bullshit, man. Don't lie to me!" Roger retorted. "I'm not going to have those guys stick me. They can never get my blood anyway!"

Chris patiently explained again the potentially serious consequences of not allowing the blood draw for this test.

Then Roger got to his real issue: his methadone dose. Addicted to heroin many years earlier, he required a daily methadone dose to suppress withdrawal symptoms.

"I don't want any of that other shit they've been giving me. I just want my methadone. I've been taking it for sixteen years, and it's all I need. I don't want any of that other crap." He made it clear that he would not allow this blood draw unless his methadone dose was adjusted to his satisfaction.

Chris again explained the importance of the blood test and tried to set aside the methadone as a separate issue. I was familiar with this conflict over the dosage. For several days Roger had complained his methadone dose was too low and attempted to make several deals to refuse treatment until the dose was raised. The residents and senior physicians attributed his bargaining attempts to the manipulative drug-seeking behavior of an addict. They did not want to be manipulated, and thus far no one had given in. Roger's methadone dosage remained at 60 mg.

As Chris continued to explain the importance of the test, his beeper went off several times. Finally, exasperated with Roger's stubbornness, he left to return his pages. I continued where he left off. As a medical student I was the only one with time to sit and convince him. For the first time during my rotation I felt useful.

I tried again to explain why it was important to know his potassium levels. I reminded him of the potential cardiac repercussions.

"I don't want to go back to the streets and that life," Roger answered. "For sixteen years I've been going to the clinic every day to get my methadone. I've been getting the same dose for sixteen years, a hundred milligrams. That's what I need. If I don't have it, I know I'm just going to go right back there." Tears welled up in his eyes. "You don't know what it's like. I used to steal and be a real bad person just to get money for the drugs. I went to jail for writing bad checks so I could get money for the drugs. That's when I knew I couldn't go on like that any longer. My life was going to hell. They told me I wouldn't make it to forty, and here I am just turned forty-one.

"I don't remember any of my time upstairs. I slept through almost all of it. I don't know what happened to me up there. But I remember when the nurses took away the anesthetic and woke me up for my birthday. My parents came, and they sang 'Happy Birthday,' " Roger said.

During his stay in the ICU Roger had celebrated his forty-first birthday. His family had visited, bringing a yellow carnation and a silver helium balloon that said "Happy Birthday" in flowery, rainbow script. He had brought these birthday gifts with him to his new room on the floor. The wilted balloon, no longer able to support itself, lay in the corner of a windowsill, bearing its wrinkled birthday tidings. The yellow carnation stood in a vase next to the balloon, looking as fresh as ever.

"Roger, you've accomplished an incredible amount. Not many people could do what you've done. No one wants you to go back to the streets and the life you led before," I said. "They have to give you all these medications to make you better. But I promise they know you need methadone and are giving you the appropriate amount. By the time you leave, everything will be all settled with your medications, and you'll be able to go back to the methadone clinic just like before."

"Aw, I don't want all this other shit they're giving me. I just want my methadone. They're not giving me the right dose! I don't need any of this other garbage. I don't want to get addicted."

"Roger, I promise that all your medicines are being carefully monitored. The doctors know what you need. You won't get addicted to this other pain medication now because you need it for real, severe pain."

"Don't you go telling me about addiction. You don't know nothing about addiction. I don't want any of this other shit. I'm going to call my methadone clinic. I've been getting the same dose for the last sixteen years, a hundred milligrams. I want the same dose now. And I don't want any of them other drugs."

I tried to bring him back to the potassium issue again. "Roger, no one expected you to live to forty, and now here you are at forty-one. We want to see you live to forty-two. That's why it's so important that you have this test."

"That's bullshit!" He picked up his phone. "Here, help me dial the methadone clinic. I'm going to ask them what my methadone dose is, and then they can tell you. I'll prove it to you."

I didn't know what to do. Was I allowed to let him call? Would I be subverting the team's efforts? I didn't want to let him do something he shouldn't out of my ignorance. I took the phone, dialed 9 for an outside line, and handed it back. "Here. I got you an outside line, but you have to dial the clinic yourself."

He made a futile effort to punch the numbers, got frustrated, and hung up the phone. "Aw, this is just bullshit!"

"Roger"—I tried once more—"the doctors are fully aware that you need methadone. But they also know what other medications you need. I'm sure they are giving you exactly what you need."

"This is just a lot of crap. I've been on methadone a hundred milligrams for sixteen years. *That's* what I need!"

"Okay, Roger." I gave up. "I'm late to a meeting right now. We'll have to talk about this later." I left frustrated. I understood his fears, and I felt for him. I was moved by his story. He had overcome so much in his life only to face this new tragedy. Yet

I was angry with him for not listening, for not choosing what I wanted him to, for not helping us help him. Roger never had his blood drawn for the potassium check.

Because Roger was so adamant about his methadone dose and because he continued to refuse treatment until the issue was resolved, Pat finally contacted the methadone clinic to inquire about his regular methadone dose. We got the faxed response later that same day. Roger was right. His regular dose was 100 mg of methadone, and we were giving him only 60 mg. We adjusted his dose the following day.

Roger also made controlling his diabetes difficult. One evening the nurses paged Chris and me to argue with him again over his nightly insulin dose. Because of his increased caloric intake to heal the wound enough for skin grafting, Roger required significantly more insulin to keep his sugars in an acceptable range than he did when he was healthy. Since his blood sugar levels had continued to run high, we had increased his insulin to obtain better control. On this evening he had refused the increased dose and requested his normal daily dose from prior to this hospitalization. The nurses had already argued with him for half an hour by the time we were called.

We arrived to find Roger looking even more pasty and subdued than usual. He wasn't feeling well that evening. But despite appearances, he was angry.

"Roger, you need the extra insulin because of all the extra calories. Your sugars have been a little high, and we need to give you the extra insulin to bring them into a better range," Chris said.

"Don't give me this crap. I want eight and eight. That's what I've always taken." He wanted eight units of NPH insulin and eight units of regular, two different forms of insulin. He needed three times that amount.

"But, Roger, we can't control your blood sugars with eight and eight. That's what you took when you were healthy. Now your body has different needs. You've been eating probably triple the calories you're used to," Chris said.

"I don't care. You don't know what you're talking about. Never in my life have I taken this much insulin. I know what I need!" he argued.

"Roger, listen to me. You need more insulin than you used to. You are taking in a lot more calories. This is *very* important to your healing!"

"Oh, man! This is bullshit! Don't tell me what I need. I'll take eight and eight and no more. Never in my life have I taken this much insulin. Eight and eight! That's my limit! No more! I don't care how high my blood sugar is!"

"But, Roger, in your situation, eight and eight just won't be enough to lower your blood sugar. You need more!"

"Eight and eight. That's it! I don't need your crap! I just don't care!"

"Okay, Roger." We finally gave up trying to convince him. We left the room and went back to talk to his nurse.

"So, what do you want me to do?" the nurse asked.

"Just give it to him anyway," Chris said.

"And if he asks how much?"

"Just tell him we're giving him what he needs," Chris said.

"And if he asks specifically how many units?"

"Just lie and tell him eight and eight. He needs this insulin."

Informed consent was a touchy issue with patients. In the modern image of medicine, supposedly all doctors subscribed to patient-centered medicine, eschewing the paternalism of previous decades. It was no longer enough to make the medical management decisions and inform the patients of the regimens they must follow; now we laid out all the alternatives and let the patients choose the plans they felt best. If the patient made a good choice, then this system worked well, and medicine felt like a true partnership.

But what if the patient made an inappropriate choice, even in the face of all the pertinent information of risks and advantages? Even though Chris had explained the issues with diabetes, was Roger making a rational choice? And how could we tell? In the end we gave Roger the twenty/twenty-four instead of the eight and eight he requested. Was that wrong? I don't know. It was too hard to watch him make an inappropriate choice about something like insulin, which had severe immediate consequences for his overall health like dehydration and even coma. I wanted to help my patients get better, and while I knew people routinely made poor health maintenance decisions at home, it was difficult to give them the same autonomy to make bad choices when they were under our jurisdiction. Maybe we were wrong about Roger. But I knew I would have felt worse if I had let him take the eight and eight that he wanted.

In reality, doctors were biased. We laid out different plans, but we chose the data to support the option we felt best in order to encourage the patient to choose what we wanted them to choose. While no longer paternalistic exactly, we continue to wield most of the power in the patient-doctor relationship. During my anesthesia elective a woman came in for knee surgery, and unlike most patients, she hadn't chosen the form of anesthesia beforehand. For her particular surgery, it was possible to choose either general anesthesia or a regional block that would anesthetize only the leg. While the anesthesiologists usually recommend the regional block because of decreased health risks, this woman had a psychiatric disorder, and they doubted she would tolerate the hours on the operating table if she were awake.

"Just explain it to her in such a way that she'll choose the general," the head anesthesiologist told the resident who would be taking responsibility for her anesthetic.

I felt so frustrated and angry every time we argued with Roger, every time he refused another treatment. I knew he feared the drugs. I knew he tried to control his diabetes the only way he knew. More important, I knew he was frustrated with his situation. Restricted to his bed and confined to one position, he had lost his autonomy. I think that for him, saying no was the only way he could regain control over what happened to his body. Yet I was angry that he fought and bargained with us. At times I felt manipulated. Why couldn't he understand what we wanted to do? Why couldn't he at least listen to our explanations? I know I wasn't the only one so exasperated with him. The team was relieved when he was finally transferred to plastics after he was healed enough for skin grafting. We got so used to arguing with him that it became too easy to fight with him out of habit. He wasn't a likable person, and the residents and I found it hard to listen to him. When he had a legitimate concern or question, it was easy to miss it among all his irrational fears and inappropriate decisions.

I had entered the wards with high expectations for rewarding relationships with my patients. I envisioned myself as an empathetic caregiver and expected myself to overcome the hurdles to establish caring relationships. While most patients were not as difficult as Eleanor or Roger, I knew I wasn't the perfect caregiver I so wanted to be.

Months later I bumped into a resident from the hospital where Roger had been a patient. When I asked about him, the resident thought for a minute.

"Yeah, that name sounds familiar," he said. He thought for a few minutes more. "He was up in the unit for a long time, and he eventually died of kidney failure. People were convinced that his father was stealing fentanyl patches from him." Fentanyl is a relative of morphine used for pain control, delivered through the skin from patches.

"One time I was up in the unit, and his father came wobbling over to me, completely drugged out of his mind and slurring his speech. He had two patches that I could see slapped on. Yeah, that guy was there for months. I don't think he ever made it home."

Too Much

I have spent my whole life attempting to know enough. I studied many hours to master the information, to earn my A. I rushed to meet and surmount every educational challenge. I set high expectations for myself. And always I realized how much more I wanted to know, how much more I needed to experience. I was intimately acquainted with the sense of inadequacy. But in the process of my futile effort to learn it all, I never thought I would believe I knew too much, had seen too much.

An eighty-four-year-old woman came into the emergency room with her daughter. The woman lay on a stretcher in the hallway, wearing red polyester stretch pants and a white nylon windbreaker despite the eighty-five-degree heat outside. Her daughter stood next to her, fidgeting with the leather trim on her western-style purse. The mother did not seem to be in pain, but she lay awkwardly on the stretcher. I looked more closely and noticed that her right leg was significantly shorter than her left and rotated outward. Chris walked by and saw me looking at her. "*That is* a hip fracture. You should remember that position; it's classic."

I was designated to talk to the women, and in the course of our conversation I learned that the mother was a healthy eighty-four-year-old. But despite her pristine medical history, I knew that a hip fracture in an elderly woman was a serious event. It could carry a life expectancy of one year.

An X ray of the woman's hip confirmed the fracture. The head of the thighbone was in its regular position inside the joint. But where the neck of the thighbone should have been was only a jagged edge, and the body of the thighbone was displaced several inches from the joint and lying next to the iliac crest at the top of the pelvis. Now I could see why her right leg was so short.

Chris and I told this mother and daughter the results of the X ray, and they received the news well, even with relief. For them, bones healed. Yet as I stood talking to them, I knew that the life expectancy of this healthy woman had been dramatically shortened from the day before. I felt sad and burdened to carry this knowledge that neither the mother nor the daughter sitting before me even suspected.

One of my classmates worked on a medical floor where a healthy sixty-five-year-old man was admitted for vague abdominal pain. Initially she was frustrated with him for complaining despite all lack of evidence for a cause for his discomfort. "I felt like, Why are you here bugging us? Go home!" she said.

But she and her team followed up on some nonspecific tests only to discover that he had widespread lymphoma. She tried to protect the man from the impending diagnosis, paging all the different consultants and reminding them not to tell the patient. She tried to control the information, the knowledge, the prognosis.

"After telling him, I felt awful. I felt that in some way we were responsible. I mean, he came in a healthy man, and if we hadn't gone mucking around, we never would have found the cancer in the first place."

Because she found the disease, took responsibility for informing the patient, and held the key to his survival in the treatments she might offer, she felt responsible for the cancer as well.

One night, shortly after dinner, the surgical trauma team was paged to the ED. We arrived at the trauma bay well in advance of the patient. Over the radio we heard that a thirty-two-year-old man was being MedFlighted in from New Hampshire after a high-speed head-on collision. The two sixteen-year-olds in the other vehicle were killed instantly in the accident. Our patient hadn't been wearing a seat belt, and now he had no sensation below the nipples. The estimated arrival was not for another ten minutes. As more people assembled in the bay, they rushed to put together the information.

"How was he injured?"

"What was that about no sensation?"

The team members donned masks and gowns and assembled around the as-yet-empty gurney.

Several minutes later the MedFlight attendants wheeled Richard through the back door of the ED. I was nervous, afraid of what I would see. Richard was a big man with dark brown hair and a mustache. He was already stripped but covered with blankets. He had a large, bloody gash on his right forearm and another on his left thigh. His right foot dangled awkwardly from his ankle. But he was strangely quiet amid the bustle of activity among those waiting to greet him.

He responded appropriately to the chief resident's questions. He spoke calmly and slowly. He was able to recall the entire accident, gave us the names of his wife and two children, and his home telephone number. He knew the date, the time, and the name of the hospital.

Then they did the neurologic examination. The neurosurgeon quickly traced up and down his entire body, first jabbing lightly with the sharp point of his calipers and then touching with the cold round tip of his tuning fork. Richard had no sensation either of pain or of temperature from three inches above his nipples down to his toes. Then I noticed the awkward position of his arms. They were curled incongruously near his face because his triceps muscles, which extend the arms, had lost their connection to the brain. He had no reflexes at his ankles or his knees. These were grim signs, indicating a very severe spinal cord injury. The residents rushed Richard to the CT scanner as soon as he was stable.

The CT scan showed a bleak picture. Richard had fractured the fifth cervical vertebra (C5) of the neck, and a shard of bone now penetrated deep into his spinal cord. The spinal cord appeared to be severed just below the sixth cervical vertebra. The eighth cervical vertebra was also crushed. A second scan to rule out bleeding into the abdomen showed three more crushed lumbar vertebrae in the back. I knew the spinal cord transection alone would cause paralysis of his upper and lower body. But each additional image, each additional injury that by itself would have been devastating, depressed me further. I watched Richard through the window of the control room as he lay motionless on the CT table, as yet unaware of what the images held.

The neurosurgeons and the residents specializing in spinal cord medicine decided to postpone surgery to perform a magnetic resonance imaging (MRI) study, which would offer a more accurate indication of the extent of the soft tissue injury to Richard's spinal cord.

Although a large retinue of consultants and specialists was actively involved in the case and followed Richard from the ED to the CT scanner and then to his MRI, my preceptor, a nurse, and I were the only ones wheeling his gurney to the radiology suite on the third floor. My preceptor asked him again about the accident.

"I was driving in my car, and I saw the cars ahead of me swerve to the right. But I couldn't see why. Then, as the car in front of me moved away, all of a sudden I saw the car coming toward me, and then it was too late."

My preceptor, Dr. Levy, was a trauma surgeon. He was young, in his late thirties, and had small children. Although a short man, he was solidly built, and his bulky frame cut an imposing figure. But in surprising contrast with his imposing form, he was quiet and extremely soft-spoken. He fixed his ice blue eyes on his patients and spoke slowly in a low voice to them as they leaned in to catch his words. His conversation was punctuated with "All right?" at the end of every thought.

Richard's family had assembled in the waiting room. My preceptor decided to discuss the preliminary results of the CT scans with them. I wanted to go with him. I wanted to meet the family and to observe their discussion with my preceptor. I wanted to understand how they felt. Because Richard was officially my preceptor's patient, I would follow him with the surgery team, but as a student I had no official role in his care. I wasn't sure if I was allowed to go talk with his family. But mostly, I was afraid. I feared their pain, their shock, and their anger.

Before he left, Dr. Levy said to me, "I hate this. I've done this so many times that I know exactly what the family will be like and how they'll respond. And I know how they'll be ten days from now too." He turned and left to meet them in the waiting room.

Just after my preceptor left, Richard was wheeled into the MRI room. The twenty of us following Richard crowded into the MRI control room, waiting for the images of his spine to appear on the computer screens. As I sat on the floor in a corner of the room waiting, my feet too painful to stand after a long day, I thought back to my first meeting with my preceptor. He specialized in trauma surgery, and I had thought his job would be so fascinating and exciting. But he didn't seem nearly as enthusiastic as I.

"Well, it's okay," he had said, "but I don't think I can do this for the long term." He hadn't explained why.

Later that same evening I had arrived home barely in time to make my aerobics class when he paged me back to the hospital for an emergency gallbladder surgery. As we began the operation, he mentioned how disappointed his children were. They had planned dinner at Chuck E Cheese's that evening. When the emergency came up, they had to cancel their evening plans.

"My wife wasn't too upset," he said. "We learned during my residency never to make plans. That way there's no disappointment."

I remembered thinking at the time, This is why he doesn't like trauma: unpredictable and long hours. He has so little time for his family.

But as I sat tucked in my corner of the MRI control room, I finally understood. My feet ached, and my dry tongue seemed to take up my entire mouth. My teeth and jaw throbbed with tension. I worried about what would happen to this man and his family. I was afraid of what these new scans would show. Now I understood.

Unexpectedly the MRI injected a slim hope into the bleak clinical picture. Although the cervical spine of the neck was grossly distorted, a small area of cord appeared to pass continuously, although stretched, through the damaged canal. Although it was unlikely to be functional, it was still intact.

With all these studies, the neurosurgeons understood the extent of his injuries and were ready to plan a surgical solution. When I left that evening, Richard was already on his way to the operating room to be fitted with a halo, a metal brace providing traction and immobility to his spine for the healing process. As I walked home, I thought about Richard. This morning, he had woken up a healthy thirty-two-year-old man. Tonight he would go to sleep a quadriplegic. This was more than I ever wanted to see, more than I ever wanted to know.

The next morning on rounds our team met Richard in the surgical intensive care unit. He was still sedated from the surgery. His right arm had been casted to the elbow, and his left leg was in a splint. The black metal bar of the halo circumscribing his head was attached to his forehead with two-inch screws. Four metal rods connected the bar encircling his forehead to a fleece-covered halter. His arms were still curled near his face.

Later that evening I went to see Richard with Dr. Levy. We found him more awake. His arms had uncurled. The casted right arm was lying by his side, and the left arm was bent and resting on his stomach.

"He has some use of his triceps now," the nurse told us. "It's not great, but he has something."

It was true. When we tested his sensation, he seemed to have some improvement as well. While he still could not feel his hands, he had some faint sensation down his arms and farther down his chest, almost to his nipples. These findings indicated that he had some function in the stretched but continuous segment of his spinal cord. I sat on the long counter and looked into Richard's room as my preceptor wrote a note in his chart. As I watched Richard, I felt the sting of tears forming behind my eyes. He might never regain full use of his arms or hands, but he would not be a complete quadriplegic.

In all the time Richard spent with our team, I never spoke with him or his family myself. I saw them only with my preceptor or the team. Because we often came by at early or late hours, I saw his wife only a few times, and I never met his children. His wife was an average-size woman, maybe a few pounds overweight. Her cot was always set up in the corner of Richard's SICU room. Her abundant shoulder-length hair surrounded her face in feathery brown layers. Her mouth was set in a thin line of worry, and her large brown eyes betrayed her sorrow and fear. I watched her lean in toward my preceptor, struggling to hear his soft, slow voice. She strained with every muscle in her body to glean every last little hope he had to offer.

After a few days on our service Richard was transferred to the neurosurgery team. I lost track of him for almost a month until I did the "Nurse for a Day" part of my surgery rotation. Richard was not the patient of the nurse I followed, but early in the morning we helped transfer him from his bed to a gurney.

The darkened room barely accommodated the gurney next to Richard's bed, and it was a tight squeeze for me to get around the corner of the bed to the other side so I could help roll Richard. I was just thinking about how poorly designed this room was when I suddenly realized why it was such a tight fit. As I tried to get around the corner, I nearly tripped over a cot I hadn't noticed. I looked over and in the shadowy corner saw his wife, whom we had clearly woken by our efforts with the gurney. As she sat in the dark, her hair was still flattened against her head from sleep. I saw tears silently course down her cheeks and drip off her chin as she watched us struggle to move her husband.

Later in the day I saw a little boy who looked almost three years old giggling and laughing as he ran down the hall. Richard's wife ran after him down the hall, scooped him up, and gave him a kiss on the cheek. Her mouth was smiling and happy, but her brown eyes were still tired and sad.

Another evening just a few days before the end of my surgery rotation, the surgical on call residents and I had managed to get through sign-out rounds and the other early-evening activities in time to make it to the cafeteria before it closed. We were just finishing dinner when we got the trauma page. We hurriedly put away our dinner trays and rushed to the ED, where we arrived several minutes before the patient. The radio report said that a man in his thirties riding a bike had been hit by a car at an intersection only a few blocks from the hospital. He wore a helmet, but there was head trauma. Witnesses said the patient lost consciousness during the accident, but he was now alert. We grabbed blue gowns and gloves in preparation for the patient's impending arrival. I stood at the foot of the bed next to my intern, trying to be unobtrusively present.

In just a few minutes the EMTs pushed through the front door of the ED and wheeled the patient into the trauma bay. The patient groaned in pain. He lay on the backboard, his neck restrained by a black collar. Rectangular yellow pads next to his ears stabilized his head, and rectangular orange pads were belted to his right leg to protect his obvious open "tib-fib," a fracture where the tibia and fibula, the shinbones, have broken and then punctured the skin. A star-shaped gash on his forehead oozed blood. His hair, saturated with blood, was sculpted into a long, narrow spike running down the center of his head. It was impossible to determine its original color.

Although in pain, the patient was alert and talking. His name was Craig, and he couldn't remember anything about the accident. He complained of pain in his left shoulder and neck. The interns quickly cut off Craig's clothes and exposed him. He was deeply tanned, with starkly contrasting white skin around his now-naked hips.

He could feel sharp pricks and sense cold temperature over his entire body, and he was able to move all his limbs. When the intern pulled on his left arm to arrange it for the neck X rays, he felt the crunch of shattered bone beneath his fingers.

Craig's neck didn't look quite right on the plain films, and after the necessary plain-film X rays were completed, I helped wheel him up to CT. He groaned in pain, with each moan regularly following the one preceding it. I was surprised by how easily I was able to distance myself from Craig's pain. The groans quickly receded into the background. Despite the horror of the man in front of me, I had little difficulty focusing on the tests he urgently needed.

We transferred Craig on his backboard from the gurney onto the sliding table of the CT scanner. We wrapped him tightly in a white sheet to prevent him from moving, and then we left the scanning room for the control room next door. We watched through the window as the table slowly moved Craig through the scanner. Over the intercom we could still hear his regular moans. We decided to finish the scans and then push his bones back in place in the scanning room afterward. One person would hold the leg at the knee, and another would pull hard on the foot, push down on the leg, and bring the bones in line. I was designated to help hold the leg for splinting.

The first scans of his head were promising. Although he had sustained a large gash, his skull was not fractured, and there was no evidence of bleeding in his brain. But we were horrified by the scans of his neck. He had crushed his first neck vertebra, C1. Then we saw C2. It was also crushed. As were C3, C4, and C5.

The spinal canal was somewhat narrowed by the movement of the bony fragments, but there did not appear to be any pressure on the cord itself. In the miraculous absence of any clinical signs of neurologic damage, it was decided that he did not need an MRI because there was no evidence of spinal cord injury. Instead he would go straight to the OR to be fitted with a halo. He also needed a more permanent surgical fixation of the tib-fib and pins to stabilize the humeral fracture in his left arm.

After the scans were completed, the three of us appointed to manage the reduction of the tib-fib went through the door into the scanning room. On the count of three, the resident and the intern pulled, and the orthopedics intern guided Craig's leg into alignment. He groaned in pain as we reset his leg, but afterward he felt much better. Once his leg was splinted, Craig went directly to the operating room for the halo to stabilize his spine and surgery on his leg and arm.

Two orthopedic surgeons I didn't recognize walked into the OR to work on Craig. One was a blond-haired, blue-eyed man with the physique of a football player. The other, a black man, pleasant and soft-spoken, rivaled the blond surgeon in bulk and girth. They took the halo kit from the nurses and rapidly assembled the bars and screws. Using local anesthetic, they screwed two screws into his skull and attached the horizontal bar of the halo. They then attached a gray bar to the head of the operating table, and after tying a ten-pound weight to the circumferential bar of the halo, they draped it over the gray bar to provide traction to hold the vertebrae in alignment during the operation. They would finish assembling the halo after the surgery was completed.

Because of the spinal fractures, the anesthesiologists, who were afraid of damaging Craig's spinal cord, could not manipulate his neck into the position of extension required to insert the breathing tube easily. Patients were normally given deep sedatives that caused them to stop breathing, at which time the anesthesiologists could insert a breathing tube with no discomfort. Since they couldn't manipulate Craig's neck, it was too risky to give such deep sedation. They worried that they wouldn't be able to establish an airway in time. Instead Craig would need an awake intubation, whereby he would forgo initial sedatives before insertion of the tube. He would be awake when they passed the tube into his trachea (windpipe).

I recognized the anesthesiologist on call. I had seen Michelle many times in the OR. She had short brown hair and vivid blue eyes above her standard blue mask. She had a strong, energetic personality that made people want to be around her. She had a great reputation among the surgeons. I often heard them talking about her in the hallways.

Five people were arranged around the operating table, and I stood a few feet back from the foot of the table watching. Michelle stood at the head of the bed. She pulled out a long black fiber-optic scope for the intubation. While watching Craig's airway through the scope, she would snake the black tube through his mouth down his throat and between his vocal cords. Then she would feed an endotracheal tube over the fiber-optic scope to place it securely in his trachea.

But almost immediately things were not right. Craig had sustained internal injuries to his tongue and airway. Michelle's view was obstructed by blood, and she couldn't guide the scope very well. Craig, because of his injuries, had not received any sedation, and he fought the tube.

"I can't breathe! I can't breathe!" he yelled. He brought his hands up and grabbed for the offending instrument. "I can't breathe!"

"Yes, you can, Craig! You can!" the nurses yelled back at him as they tried to restrain his arms.

Michelle tried again.

"I can't breathe!" He gagged against the long neck of the scope. "I can't breathe!"

"Yes, you can! Look, you're talking to us!"

The nurses fought with him to keep his hands away from Michelle and the scope. They yelled at me to come and hold down his right hand.

I leaned with all my weight to keep his arm safely by his side and out of reach of the anesthesiologist. Still, Michelle could not guide the scope into his trachea.

Michelle called in another anesthesiologist, and the surgeons began to discuss the possibility of creating a surgical airway: making an incision through the neck directly into the trachea to insert a breathing tube. Craig coughed and gasped for air now that his airway was rid of the offending instrument. The clock said twelve thirty-five. I was tired and tense standing in my position at the table. I still held Craig's hand, trying to find comfort for both him and me. In the brightly lit OR full of activity and tension, the atmosphere was surreal. There were no cues to the time of day. It could have just as easily been the middle of the afternoon. Then I realized this was my recurring dream from anatomy class in first year. In the dream I held my cadaver's hand to comfort her as my partners performed some painful procedure. Now, holding Craig's hand in the middle of this surreal night, I lived that dream.

The new anesthesiologist, an older woman, passed the tube easily through his nose and into the vicinity of his larynx (voice box). As the tube touched his larynx, Craig began to spasm. Every muscle in his body tightened, and his torso seemed to grow smaller as he bucked on the table. Still holding his hand, I felt the rhythmic muscular contractions, and through my hips leaning on the table, I felt the vibration of each impact. I watched his neck, no longer restrained in a stiff collar, move up and down in concert with the rest of his body with each spasm. The oxygen saturation in his blood steadily fell. Normal was at 98 to 100 percent, and I first noticed Craig's sats at 80 percent when the alarm went off. I watched as the green numbers rapidly fell to 70 to 60 to 55. The anesthesiologists decided to do a last-ditch blind intubation before attempting a surgical airway. The older anesthesiologist tried once. She inserted the tube and gave Craig a breath while Michelle listened with a stethoscope at his lungs.

"No good! It's in his esophagus."

His O2 sats dropped further: 50, 40.

"Okay," the older anesthesiologist said, "we're going to have to do a surgical airway."

Pat and Dave were on call that night. They had already retrieved the kit to create a surgical airway. "It happens about once a year in a hospital like this," the senior resident had told me.

Dave calmly made a small incision into the neck and opened a small hole into the trachea. He then inserted the curved tip of the breathing tube directly inside Craig's trachea and connected them to the ventilator tubing. Finally, Craig got a breath. The O2 sats began to climb again and soon reached his previous levels of 100 percent.

I worried that the movement of his neck during the spasms had damaged the spinal cord, and his brain had been without oxygen for a few minutes until they created the surgical airway. But because of the anesthetic, there was no way to tell what damage, if any, had occurred until the morning, when the drugs wore off.

No one needed my help, so I left the OR and went to get some sleep. Visions of Craig's awake intubation played in my mind. Of all the blood and pain I had seen in my weeks on the trauma team, this awake intubation was without a doubt the most horrible event I witnessed.

We met Craig the next morning in the ICU. He was still covered with the crusted blood from the night before. His operation had gone very well, and he didn't seem to have suffered any neurologic damage during his spasms the night before. Because of the breathing tube, he couldn't talk, but he tried to communicate with us by forming words with his lips.

After the surgery Craig's tongue became so swollen that it would have obstructed his airway, so he continued to require the neck tube to breathe. Several days later the tongue swelling went down enough to extubate him, and he was discharged from the ICU.

A few days later I saw Craig farther down the hallway walking on crutches. He moved clumsily a few steps away from his doorway before I turned the corner and lost sight of him. I was thrilled to see him walking; he had just barely escaped becoming quadriplegic only a week before.

But that evening at dinnertime Chris and I were paged to see Craig. He sat in his bed, and his mother was with him, holding a forkful of diced carrots.

"How can I eat this?" Craig said angrily, and gestured to the plate on his tray. "I can't swallow *that*! What are you doing, giving me this stuff? And rice—how do you expect me to eat that? I need applesauce and soup. Not this stuff!"

Chris explained that we were concerned he would have less control over thin liquids and might inadvertently inhale them into his lungs. The resulting pneumonia could be deadly.

"And the walking. I can't walk like this!" Craig complained. "I can't live like this! How do you expect me to get around if I can't walk?"

I left Craig's room frustrated. I knew he was angry at his situation and his accident. Wearing a halo for three months was not a pleasant prospect. But I knew how bad he could have been. I thought of Richard. Craig was lucky to be alive and walking at all. When he lay bucking on the operating table with an oxygen saturation of 6 percent, his prognosis wasn't so optimistic. But of course Craig had no concept of what could have been. His only basis of comparison was with how he used to be. An inability to talk and swallow and breathe and walk easily would be devastating, even in the short term. But I wished he could share my joy and excitement.

I left the team for another rotation before Craig was discharged from the hospital. I never found out what happened to him. I was happy to leave the trauma team at the end of the month. For the first time I understood what it was like to care for hospitalized patients and become emotionally invested in their care. The pain was more overwhelming than I ever imagined it would be.

Obstetrics and Gynecology

Beth sat silently on the birthing bed, the back raised high and the foot dropped low to create a soft chair. Warm yellow light suffused the late-afternoon room. The acrid scent of amniotic fluid perfused the air, and the constant beep-beep-beep of the baby's heartbeat, translated to us through the monitor, blanketed the room. Beth's wavy auburn hair was pulled back into a ponytail; loose strands and bangs shaped her pale face. Her eyes squeezed shut, and her mouth pressed into a thin, mauve line. She looked down and to the left, slowly exhaling "Tssssssssssssssss..." with each contraction. Her johnny hospital gown pulled slightly over her taut, swollen belly, distorted by the two pancake-shaped receivers of the fetal heart rate and contraction monitors. The flimsy cloth of the johnny was carelessly pulled up, exposing the darkness of her pubic hair with its delicate tendrils formed by now-dried amniotic fluid. There were five of us in the room, with Beth as the focus. Her husband sat beside her, and occasionally she grasped his hand during a contraction, her knuckles white from squeezing and her arm trembling slightly with the effort. The nurse, the nursing student, and I sat off to her left, next to all the monitors, eager to fill Beth's needs and to chart her progress. This long afternoon as the light slowly faded, we waited together for Beth's baby to come. I was struck by the solitude of the endeavor, each of us lost in our own waiting.

The baby finally came in the dimness of an early-winter evening. I stood and held one of Beth's legs against my hip while her husband held the other, as Beth forced the contents of her bowel and uterus onto the green draped table. The pungent scent of amniotic fluid, blood, and stool suffused the room. Olivia broke the tense waiting and pain with her first cries. Overpowering joy welled up inside me and threatened smarting tears. Beth shared a controlled joy as she held her third child and first daughter in her arms.

I spent three weeks working on the labor and delivery floor as part of my obstetrics rotation. The labor and delivery ward was on the top floor of the hospital, and each birthing room was large and had a small alcove with windows. On one side of the hospital, rooms 8 through 14 looked out onto the irregular gray concrete rooftops of the other buildings composing the hospital, but on the other side, rooms 1 through 7 overlooked the multicolored fall hills and neighborhoods. I watched many a father escape to the relative peace and solitude of the alcove when his wife's or girlfriend's labor became too intense. They made me angry, those men sitting there or trying to sleep, leaving the nursing staff to support their wives through painful labor. The birthing bed itself was positioned in the center of the room, flanked on the sides by fetal and maternal monitoring equipment and IV poles. A small baby table stood near the door, just a few feet from the birthing bed. The tabletop was padded with soft pink and blue blankets, and a large light above could be turned on to warm the newborn infants. Before each baby was born, the table was stocked with a diaper, a cotton baby shirt, and a piece of undyed cotton tubing knotted off at one end to make a miniature cap.

I spent most of my time at the nurses' station at the center of the ward. A large wipe board that tracked all the women on the floor through their labor hung on the wall behind the counter. Each woman's room number, name, week of gestation, number of previous deliveries, and stage of cervical dilation were recorded and then updated throughout her labor. At the beginning and end of each shift, the residents collected here to discuss each woman's situation and her likely course over the next few hours.

The two operating rooms were directly across from the nurses' station. All cesarean sections, planned and emergency, and most tubal ligations were performed in these operating rooms, which looked essentially like the regular operating rooms in the other areas of the hospital except for the baby table at the far right of the room.

I watched and even assisted on cesarean sections during my rotation, but I never got used to them. They were bloody and difficult to follow. The first part went very quickly. The obstetrician started with an incision just above and parallel to the pubic hair, located the bladder, separated the two halves of the rectus abdominis stomach muscle, and grasped the uterus. After a quick incision to the lowest part of the uterus, the obstetrician reached her gloved hand deep into the hole she had created and pulled out the head of the baby. The rest of the baby rapidly followed, and the obstetrician cut the cord and passed the baby to the waiting nurses for cleaning, warming, and dressing. After delivery of the placenta, however, the rest of the operation could take up to an hour. The rapidly made uterine incision was often difficult to repair, and the continued postpartum uterine bleeding made the situation even more complicated.

The mothers having C-sections received spinal anesthetics, leaving them awake and alert but dulled to the pain so they could participate in the birth. A sterile green curtain at the level of the diaphragm prevented the mothers from seeing the obstetricians incise their abdomens. The fathers were allowed to come into the operating room and sit with their wives at the head of the operating table after the initial incision had been made. Too many fathers passed out to be allowed in from the outset, but they were always present for the birth. Quite a few proud fathers captured the moment on video; it always seemed a little macabre to me.

I saw many babies born during my month on the service and even caught a few by myself when they came so quickly that the senior physicians had not yet scrubbed in. I especially enjoyed Evelyn's baby. Her delivery was a family affair, with her twelve-year-old daughter, her husband, and her parents in the birthing room to share the moment. Her daughter was bored; she had sat in that room for seven hours since being taken out of school after an English test that morning. She read us all our fashion horoscopes from *Young Miss* magazine to while away the hours of waiting. It turned out that she and I shared the same sign, Aquarius.

" 'You are soon going to be able to call that special boy in your life your boyfriend.' " She considered for a moment. "Well, that must be for you because it sure isn't mine," she said.

I guided her hand over her mother's uterus to help her feel the contractions, explained what would happen, and, when the birth was imminent; pointed out the first tufts of hair visible between her mother's labia as the baby traversed its slow path through the birth canal.

But there were painful, sad births as well. Amalia was nineteen, having her second baby. She had gone into preterm labor a month before and had been given medication to stop the contractions. Since she was recalcitrant to taking her medication, however, she was almost a nightly visitor on the labor and delivery floor. The day I met her, she was upset because she had just come from questioning by the Department of Social Services (DSS), which investigates cases possibly caused by neglect or abuse. Her three-year-old daughter had taken Amalia's medication to prevent labor after she left it open on the kitchen table.

"I'm not a bad mother. I didn't mean for her to take it," Amalia drawled. "I just turned my back for a minute, and she took my medicine. It was an accident."

When she found out she was really in labor this time, she called her eighteen-year-old boyfriend on the phone. "Baby, it's coming, Baby! The doctor said the baby is coming, Baby! I'm excited to meet it finally."

But after her phone call she was nearly in tears. "He's not coming," she told me. "He's scared. I don't know why he's scared now. He's come before."

In the end her boyfriend came shortly after midnight. But three hours later they had to go home again. Amalia's labor was another false call.

There were several ridiculous births during my month on service as well. One night an overweight Haitian woman came to the floor in labor with her fifth child. After she got her epidural anesthetic, she complained that she was too tired to have this baby and sick of the pain. She just wouldn't do it, she said. Just as her cervix became fully dilated and she was ready to push, she gave up. She actually crossed her legs in an effort to prevent us from doing any more exams and to force that baby back into her womb. Much as we and her friend encouraged her and then yelled at her, she refused to help push that baby out. The baby was delivered solely on the effort of her spontaneously contracting uterus, rescued by the worn elasticity of her tissues from birthing four other babies. Her body easily stretched to accommodate his tiny form.

Obstetrics is a wonderful profession because physicians care for young, generally healthy women during a very happy time in their lives. Yet despite the joy of birth, I found it hard to watch women in labor. The messiness and smelliness of labor disgusted me. I was appalled by women defecating on the birthing bed as they struggled to force their babies into the world. I was horrified by the amount of bleeding and embarrassed by the sheer number of pelvic exams required.

My strong reaction to the grossness involved in delivery surprised me. This was not my first experience with the distastefulness of bodily functions. I performed rectal exams regularly. I collected urine to check for infection. I examined phlegm. I spent three weeks on the trauma service. All these experiences seemed routine rather than ugly to me. How could I have been so revolted by women in labor?

It was denial, I eventually decided. Being in medicine demanded a supreme amount of denial. It would be impossible to survive the hospital if I constantly worried that I or my family members would develop the diseases my patients suffered. While people always joked about the hypochondriac medical student who "contracts" every disease he studies, my classmates and I had assumed a certain amount of immunity by the time we hit the wards. We saw people sick and dying, both young and old. We saw healthy people mutilated and destroyed by accidents. But we still had to ride our bikes, drive our cars, and cross the streets. When Carlos bought renter's insurance, and the agent tried to sell him a life insurance policy, Carlos remarked, "Life insurance? What would I do with life insurance?" It was denial.

But watching women in labor, I had no refuge in denial. I hoped I would have a baby someday. I would be in the same situation as these women I watched. When I stood in room after room, waiting for the babies to be delivered, not only did I follow the clinical progression and study the medical decision making, but I also assessed what I would and would not want for myself. The grossness bothered me because I knew I would be in labor too someday.

Carlos did not share my response at all. "What are you talking about?" he said to me. "It's natural. There's nothing to be ashamed of."

I think it was different for him as a man. He observed empathically without the knowledge that he would one day be in the same position. It was easier for him to miss the humiliation and unpleasantness.

During third year we continued to meet once a week for Patient-Doctor. Instead of hands-on clinical teaching, the course was now a two-hour lecture on the principles of medical management followed by a discussion session in our small groups. Whereas Patient-Doctor was the highlight of our course work during the first two years, now that we were on the wards, it seemed more of an afterthought. One Tuesday afternoon I walked home with Alyssa after Patient-Doctor. She had just finished her gynecology rotation, and we compared experiences. Her husband was also a Harvard medical student two years ahead of us, so their relationship was in some ways parallel to Carlos's and mine.

"You know, it was the same with us. I didn't like labor and delivery at all. It really was gross—all the blood and amniotic fluid. One day I was in the emergency room, and a woman came in who, from the history, sounded like she was having a miscarriage. When I asked her if she was passing any blood clots, she pointed to her black leather handbag sitting on the floor in the corner of the examining room and told me that she had collected the tissue she was passing in a little plastic bag. I went over to the purse and pulled out the plastic bag, and inside there was a tiny fetus, about three inches long. I was horrified. But Alex didn't have the same response at all. He couldn't understand at all what I was so appalled by in ob-gyn. I think it's just that it's not at all about him, whereas for me that could be me there on the birthing table."

Obliterating the barrier of denial in my experiences with these women introduced me to a degree of intimacy in the patient-doctor relationship I was unprepared to confront. Whereas before, my patients were simply sick people, suddenly I saw myself in all these women.

By the time I came into my ob-gyn rotation, I had already

lived three months of hospital life, and the novelty of working in the hospital had worn off. The first three months of my surgery rotation had been exhilarating and exciting, but as fall quickly erased the warm comfort of long summer days, I confronted a new and unsettling layer of self-doubt.

The TABs

In ob-gyn, lingo took the place of language. Three-letter acronyms dominated conversations, and it was possible to read an entire note without coming across even one whole word. A woman might need a TAH, LSO, BSO, or a TVH. A woman might have PID if she had CMT and D/C. Or she might need a D & C for a missed SAB, after which she still might have retained POCs. But Alice, Tonie, and Danette had none of these things. They needed a TAB.

TAB stands for "therapeutic abortion." This was not a required part of my gynecology rotation, so I specifically asked my resident if I could attend. After spending years thinking and arguing about this procedure, I wanted to know what it was really like.

The TAB procedure actually spanned two days. Each woman arrived at the outpatient women's health center on a Tuesday morning. She first met with a social worker, who offered the opportunity to discuss her decision to terminate her pregnancy. The social worker assessed the woman's comfort with her decision and ensured that she had sufficient support. The social worker also questioned her about her plans for contraception in the future. The woman next visited a nurse, who took a detailed medical history and explained the procedure. The nurse also again addressed her plans for future contraception. Finally the woman met with the resident who, under the supervision of a senior physician, performed the procedure the following morning. The resident obtained informed consent from the woman and then inserted laminaria, which are small seaweed toothpicks that dilate the cervix overnight to make room for the evacuation tube during the procedure. Once the cervix began to dilate, the woman could no longer change her mind; the fetus would not survive. The resident also wrote a prescription for the form of birth control that the woman requested during the visit. The following morning the woman arrived at the outpatient surgery center for the termination.

Tonie, a tall black woman, was first on our list that morning. I went with her to all three visits. Tonie wore her short, straightened hair fastened with a metal clip at her right temple. She was dressed in a bold black, red, and white striped T-shirt with red leggings and black ankle boots. At twenty-four, she was exactly my age.

This was Tonie's second pregnancy. She had had a son, now seven, with her high school boyfriend. "I just don't want to do that again," she told the social worker. "I was young and naive then. I thought, Okay, we love each other. We'll get married. I thought it would all work out in the end. But now I'm more mature. I promised myself that I would never have another baby unless I was married, you know what I'm saying?"

Tonie felt ambivalent about terminating her pregnancy. Her boyfriend was willing to help her raise the baby, but he said he would be equally supportive of her choice to have an abortion.

"You know, I just wish he would make a stand on one side or the other, 'cause it's like he doesn't care, really. It's all my choice, you know what I'm saying?

"I was thinking about keeping it because my boyfriend, he don't have any kids, and he'd like to be a father. But I don't want to have another kid now. I just got enough money to finish my last year of college, and I want to be a lawyer. I know if I have this baby now, I'll have to take more time off. And maybe my boyfriend and I will decide in a year that we don't want to be together anymore and go our separate ways. What will I do then? That's what happened to me when I was young," Tonie said.

After our meeting with the social worker, Tonie and I moved to the office next door for the session with Janice, the nurse. Janice sat behind her desk while Tonie and I sat in chairs side by side across from her. With her short gray-blond hair, soft cotton clothing, and bright red cheeks, Janice exuded an aura of comfort. Her matter-of-fact demeanor and absolute lack of judgment offered safety and protection. Even if Tonie was conflicted about her choice or ashamed of her situation, to Janice this was normal and routine.

I sat with Tonie as she provided her medical history and her insurance information. I listened with Tonie as Janice explained what would happen in her last appointment of this morning and when she returned the following morning. Tonie had a few questions; Janice answered some and deferred the rest to the resident.

I had just gotten my period the day before. Sitting in that office, in my chair, side by side with Tonie listening to Janice, I had never before felt so acutely aware of my bleeding, so conscious of my unfulfilled reproductive potential.

Finally, we were ready for the last visit with Sam, the resident. By this time most of Tonie's questions about the procedure had been answered. Sam went over much of the same information again, and Tonie assured him that she was prepared to undergo the procedure. Sam then wrote her a prescription for the Pill, and Tonie signed the necessary consent forms. Sam and I left so Tonie could change for the physical exam.

When we returned, Tonie had shed her bold clothes and lay in a johnny on the examining table. We prepared to do the trans-vaginal ultrasound to determine what size instruments he would need to use for the procedure. Sam switched on the ultrasound machine and positioned the monitor so it faced toward us and pointed away from Tonie. He took the long, narrow probe and squeezed green gel on the black tip. He opened an orange-wrapped Trojan condom and rolled it down over the probe. He inserted the prepared probe into Tonie's vagina, and she squirmed with the intrusion. In just a few moments the kaleidoscope of shadows on the screen merged into the form of a baby. The baby was actively rolling and turning inside Tonie's womb. As Sam rotated the probe to explore all the dimensions of the baby's body, we found a rhythmic flicker in its chest.

"That's the heart beating," Sam explained to me quietly.

As Sam focused intently on the screen, struggling to find the best images to achieve the most accurate information, I noticed Tonie had craned her neck to get a view of the image on the screen. She lay quietly watching this unwanted life kick and squirm within her. I don't think Sam noticed she was looking. As she watched the screen, I saw a tear etch its path into her cheek from the corner of her eye to her ear.

When Sam and I finished with Tonie, Janice met us outside in the hallway. The next patient was waiting for us already, she said. "Be careful with her," she warned us. "Danette is sixteen years old and sixteen weeks pregnant. She has a one-year-old baby at home, and this is her first abortion." Janice sighed. "It's just a terrible situation."

We met Danette in the waiting room. The minute she saw us with Janice, she stood up and walked toward us. She had been sitting next to another young black girl. They both wore oversize plaid flannel shirts untucked over baggy jeans. Danette was strik-ingly beautiful. Her hair was pulled back into a short ponytail. She had opalescent creamy brown skin and delicate features. Very tall, she walked serenely ahead of us to the examining room with a sense of composure and dignity I had never seen in a sixteen-year-old. Despite her innocent beauty, she had the self-restraint and aloofness of someone who had been hurt deeply.

Danette answered our questions quietly with sparse language. She had obtained consent for the procedure from her legal guardian. She lived with her aunt and her one-year-old daughter. She had not been using any form of contraception since the birth of her first child. At one point during the interview Sam was called out to answer a page. I sat in the small closet of a room with Danette. I wanted to get a sense of her home situation, an idea of what her next twenty-four hours would be like. I wanted to know how she felt about her situation. I wanted to forge a connection. But she resisted all my efforts. She remained closed into herself, answering my questions with only the most minimal responses. She seemed resigned to her fate.

When it came time for the ultrasound and laminaria insertion, Danette changed obediently, without a word. She lay quietly on the examining table, fingers interlaced over her breastbone. She wore a thin gold chain draped around her right wrist. Just above the bracelet was a two-inch paisley-shaped scar, mottled brown against the smoothness of the rest of her skin. Her rich brown eyes focused intently on the ceiling. I saw the tension she exerted, forcing her lips to remain still and neutral, defying the temptation to cry out against her situation.

Sam did a pelvic exam. He showed me the blueness of her cervix, a sign of venous pooling from the enlarging uterus. He palpated her growing uterus, just beginning to swell beneath her flat belly. Because her fetus was so developed, Sam did a transabdominal ultrasound, sparing her the vaginal probe. Her sixteen-week-old fetus easily made itself visible, distinct from the other shadows in her abdomen. The shadows of the beating heart did not require explanation. While Sam attempted to measure the fetus, Danette quietly turned to look at the image on the screen. While younger fetuses were easily lost in the shadows, unidentifiable to the untrained eye, Danette's baby was easily recognizable.

This time Sam noticed her looking. "You can watch the screen if you want. Some women like to," he told her. "But oftentimes they feel it's too upsetting. Anyway, you can watch if you want."

Danette did not turn away. A small quiver escaped her controlled face, and a tear spilled down from her eye. I went to touch her scarred arm, to offer comfort. While she did not flinch beneath my hand or make an effort to remove it, she did not touch me back. Her arm lay inert beneath my fingers. I quickly took my hand away and moved back toward Sam at the foot of the examining table.

Because the fetus was so large, Danette required extra laminaria to make way for larger instruments. I watched her struggle to maintain her impassive face, but a grimace escaped periodically as the cramping from the insertion set in. When the exam was finally over, she quickly got up off the table, her relief palpable. She re-dressed in her oversize flannel shirt and baggy jeans and quietly left our office.

Alice was the last woman scheduled for the morning. Sam and I looked at her chart in the hallway outside the examining room. At the age of thirty-eight, she had been pregnant twelve times, borne five children, and had six TABs. By her last menstrual period (LMP), she was estimated to be eight weeks pregnant. "Can we say tubal ligation?" Sam said, joking about surgical sterilization.

Alice was an obese Hispanic woman, with dingy brown skin. Her overpermed hair stood away from her face in an artificially blond frizz. Bright pink lipstick coated her thick lips, constantly open to reveal a thin glimpse of gray teeth. She wore a navy blue hooded sweatshirt and too-tight jeans. Her nose and eyes were the mottled red of recently shed tears. Alice already knew the routine. She answered the questions easily with the wavery voice of barely restrained emotion. She signed the necessary forms and changed quickly.

When we returned, she sat on the edge of the examining table wearing the hospital gown, tube socks with three red stripes at the calf, and her sneakers.

"Did I need to take off my shoes and socks too?" she asked.

After removing her shoes, we guided her stockinged feet into the stirrups. Her knees shook as we tried to relax her to insert the ultrasound probe. Alice turned her head away from the screen as Sam dated her fetus. As he began to insert the laminaria, Alice broke into shaking sobs. Her inner thighs and labia rocked in concert with her tears.

After ensuring she was not in pain from the procedure, Sam reassured her. "I know this is difficult, but just bear with me. This will all be over soon."

I handed her a box of tissues. Sam had to wait for each sob to subside before he could insert another laminarium. Finally the procedure was over. Alice sat up, her eyes and nose now fiery red, grasping a tear-soaked tissue in her right hand.

"Are you going to be okay until tomorrow?" Sam asked her. She assured him she would.

"Is someone here with you today?" Sam asked. She said she was there alone, but when she called, her boyfriend would come and pick her up. With that, Alice squeezed back into her too-tight jeans and navy blue sweatshirt and left our office.

By the time I arrived at the outpatient surgery area the next morning, Alice was already gowned and lying on a stretcher in the holding area, awaiting her procedure. Her eyes and nose still had the look of recently shed tears, and her frizzy hair puffed out the blue bouffant surgical cap required in the OR. Her lips were stripped of their pink lipstick.

Alice recognized me immediately. Sam arrived soon after me and explained again what would happen that morning. A few tears overflowed and slipped out the corner of her eyes as she listened. Then it was my job to start her IV for the procedure. Sam left to find Dr. Elliott, the senior physician. Although I was comfortable with the steps of inserting the IV, I was inexperienced. I worried that a miss would compound her already substantial anxiety, and I feared her obesity would obscure her veins.

I wound the elastic tourniquet tightly around her upper arm and was pleased to see a large, straight vein standing out in her hand. I cleaned her hand with an alcohol wipe and was relieved to feel the IV catheter slip smoothly and easily into the vein. I attached the tube of IV fluid solution and watched it run freely into her arm. As I talked Alice through the IV placement, she became calmer. Her liquid brown eyes no longer threatened imminent tears.

We still waited for Sam to return with Dr. Elliott, so I asked how her morning had been. Alice burst into tears. "My boyfriend didn't even talk to me about it this morning. It's like it's my problem, and I have to deal with it by myself. He didn't say one word to me. He just dropped me off here this morning and told me if I called later, he would probably be able to pick me up." Her face flamed with tears. I held her shoulder as it shook.

"I'm sorry your boyfriend wasn't more understanding," I said.

Then Dr. Elliott finally arrived. "Don't cry. It's going to be all right," he told her.

I still held her shoulder. Alice said she was okay, and we wheeled her stretcher back to the OR. She stopped crying during the trip and was only snuffling by the time we reached the OR. The nurses arranged her on the narrow operating table and put her legs in the stirrups. Dr. Elliott handed me a pair of gloves and a packet of surgilube and told me to remove the laminaria we had inserted the previous morning.

"Alice, you're going to feel my fingers touching you," I said to her. I touched her inner thighs. "I need to remove the laminaria we put in yesterday," I told her. I reached my fingers into her vagina and groped for the laminaria. I retrieved them easily.

As Dr. Elliott and Sam collected the necessary instruments and set up the evacuation machine, I went to the side table to read Alice's chart.

"Dr. Rothman...Dr. Rothman...Ellen?" Finally I realized someone was talking to me. One of the nurses called me over.

"She doesn't want you to leave her." I returned to the head of the operating table. A blue sterile drape now separated Alice's head and arms from the rest of her body.

"I was afraid you left me," Alice said.

"No, I'm not going anywhere," I told her. I reached for her hand lying outstretched in the standard crucifix position, and she grasped my hand tightly. "They're getting ready to start now, but you're going to be okay," I told her.

"In just a few minutes you're going to hear a lot of noise from one of the machines. You shouldn't worry," the nurse said.

Dr. Elliott turned on the evacuation machine, and the loud, constant *rurr* of the machine and the suck of air through the plastic tubing filled the room. Sam inserted the tubing through Alice's vagina into her uterus. Alice began to cry again. The machine gasped as Sam closed off the external valve and began sucking the contents out of Alice's uterus.

"You're okay, Alice. Everything is okay," I told her.

"This won't last very long," the nurse added.

The evacuation machine whistled as Sam momentarily opened the external valve. Then he closed it again and continued sucking from Alice's uterus. In another few minutes Sam was done. Dr. Elliott turned off the machine. Sam and Dr. Elliott evaluated the status of her uterus.

"We're all finished," Dr. Elliott called over to our side of the operating table.

As the nurses busily began to prepare Alice for the recovery room, the physician sent Sam and me out to "float the membranes." I promised Alice I would see her in the recovery room later in the morning.

The evacuation machine had collected a few ounces of blood and tissue. Dr. Elliott strained the contents of Alice's uterus through a linen-colored loosely knit sock. Sam and I took the unrecognizable lump of tissue collected in the sock to the sink in the hallway. I held the sock open as Sam ran water through it. It drained pink into the sink basin.

"We have to wash away all the blood," he explained to me. When he was satisfied, I filled a clear plastic bucket with water, and he dumped the soggy bits of tissue in. Most of the red tissue sank to the bottom of the bowl. But a thin pink, membranous circle of tissue remained floating serenely in the water, reminding me of a jellyfish.

"That's the amniotic sac," Sam explained. "We have to make sure we see it so we can be certain we have disrupted the pregnancy."

Satisfied, Sam strained the fluid in the bowl back through the sock, and we retrieved the tissue. He returned the tissue to Dr. Elliott, presumably for disposal, and sent me out to the holding area to start the IV in the next patient.

Danette, the sixteen-year-old, sat in a blue recliner with wheels in the holding area. She had already changed into her hospital gown but was not yet wearing the blue surgical cap. Alone, she waited passively for the procedure to begin. I asked her how she was doing.

"Fine," she said.

How had her morning been? I asked as I set up the materials for the IV placement.

"Fine," she said again.

I asked her if she had ever had an IV before.

"You mean besides my first birth?" she asked.

I quickly placed the tourniquet on her arm and examined her hand for appropriate veins. I picked one on her wrist. I injected a small amount of lidocaine, and just as I inserted the needle into the vein, Dr. Elliott and Sam came up behind me.

"Why did you choose that vein? Look at these big ones here." He pointed to her elbow. I had found the vein easily, but now I hit resistance as I tried to pass the IV catheter deeper into her vein.

"This is no good!" Dr. Elliott yelled. "Look, this vein is kapooey! Pull it out!"

I pulled out the catheter.

"No, not like that! You're going to give her a hematoma! Look, now she's going to have a hematoma!" he yelled.

Danette covered her eyes with her free right hand. Sam reached in to apply a pressure dressing. The three of us crowded over Danette in her chair. I still had the needle in my hand, and as I rushed to get it out of the way, I nearly stuck Sam as he reached in with the gauze. Finally, the three of us sorted ourselves out. I taped the pressure dressing, and Dr. Elliott placed the IV quickly and easily into the elbow vessel he preferred.

I was angry that Dr. Elliott had yelled at me in front of Danette. I didn't particularly care about my humiliation, but I worried about how Danette perceived the incident. I'm sure she had no idea that a hematoma was a bruise, and she must have been frightened that something I had done to her warranted this much furor. Missing an IV, while not pleasant for the patient, was not devastating either. It happened to everyone, from the most inexperienced student to the most experienced nurse. But this IV incident must have been traumatizing for Danette. This was an unhappy prelude to an emotionally difficult procedure.

We wheeled Danette into the OR in her blue recliner. When we reached the OR, she got up out of the chair by herself and lay down on the narrow operating table. Although she wanted to leave her arms crossed over her belly, the nurse arranged her arms outstretched in the crucifix position and then placed her feet in the stirrups. Dr. Elliott handed me gloves and the surgilube and asked me to remove the laminaria. After the IV incident I felt uncomfortable touching Danette again. But I successfully retrieved the laminaria. As Danette lay on the table, I watched her lips twitch while she valiantly struggled to maintain her composure. But as soon as Dr. Elliott turned on the evacuation machine, she burst into sobs.

"Are you in pain?" the nurse asked.

"I'm scared!" Danette sobbed. She cried the unabashed, unrestrained tears of a little girl.

I stood at the foot of the table with Dr. Elliott and Sam. After inserting a speculum to see the cervix, Sam inserted a series of metal rods, each a few millimeters larger than the last to ensure that Danette's cervix was sufficiently dilated to accommodate the evacuation tube. When both Sam and Dr. Elliott were satisfied,

Sam inserted the evacuation tube and Dr. Elliott turned on the suction. First, the tube filled with clear fluid.

"That's the amniotic fluid," Sam told me. Within seconds the clear fluid gave way to red. In a few more moments Sam encountered resistance in the evacuation.

"Take the macerators and crush the calvarium," Dr. Elliott told him. Sam removed the suction tube and manipulated inside Danette's uterus with the macerator, a long metal instrument. He crushed the fetal skull (calvarium), pulling out the soft pieces one by one. Sam then reinserted the evacuation tube and continued to empty the contents of Danette's uterus.

"How did you know the calvarium would be there?" Sam asked.

"Because I saw the white matter of the brain in the tube," Dr. Elliott replied. Within a few more moments the machine was finished evacuating Danette's uterus. Dr. Elliott turned off the machine, and Sam combed each wall of her uterus with a curette to ensure that nothing remained within. We did not have to "float her membranes" because we had seen the amniotic fluid, proving the sac had broken.

Afterward the nurse put a sanitary pad and a pair of net underpants on Danette. Still sniffling, she got up by herself and moved back to the blue recliner. Sam and I rolled her to the recovery room. I waved to Alice as we passed by. Alice sat by herself in another blue recliner. She smiled and waved back. We deposited Danette in a slot on the other side of the nurses' station and rushed back to the holding area to meet the next patient.

When Sam and I returned to the holding area, Tonie, the twenty-four-year-old black woman, sat in a blue recliner waiting for us. She too waited alone. Sam left to take care of some business, and I decided to wait for Dr. Elliott to place the IV. I was eager to hear how the last twenty-four hours had been for Tonie. Her boyfriend, she said, had been very good to her. He had dropped her off this morning and planned to pick her up after the procedure. She felt a little nervous earlier but was anxious to have the procedure over with.

Dr. Elliott and Sam returned quickly, and Dr. Elliott placed the IV easily. We wheeled Tonie back to the OR in her blue chair. She quickly arranged herself on the operating table and helped the nurse lift her feet into the stirrups. I retrieved the laminaria. Sam inserted the suction tube. Tonie lay quietly as Dr. Elliott turned the machine on, and the now-familiar hum and suck dominated the room.

Within several minutes the evacuation portion of the procedure was complete. Tonie continued to lie quietly as Sam inserted the curette to ensure that all contents of her uterus had been evacuated.

"Do you want to feel?" Sam asked me.

I switched seats with him, and he held my hand as he showed me how far and how vigorously to scrape the instrument against the walls of her uterus. I felt the grittiness of an evacuated uterus, as if I were rubbing a spoon over coarse sandpaper.

As the nurses helped Tonie into her sanitary pad and underpants, Dr. Elliott, Sam, and I examined the tissue we retrieved from the evacuation machine. Dr. Elliott spread the tissue onto a cardboard tray. We easily discerned the fetal parts.

"This is the rib cage, and there is an attached arm," Dr. Elliott said. "And this is a leg here," he said. The pieces looked as though they were formed from the same semirigid flesh-colored plastic of the Barbies I played with as a child.

"Don't be shy. You can touch them," Dr. Elliott said to me.

I reached out with my gloved hand and touched the miniature ribs and the tiny arm. I saw the attached neck, which I hadn't noticed before. I touched the foot lying close by.

After wheeling Tonie to her spot in the recovery room, we went back to check on Alice and Danette. Alice had already left, and Danette, still alone, rested quietly in her chair. She felt fine, she told us.

Misery on the Wards

"I got totally slammed," George said. George was a returning M.D.-Ph.D. student on his first rotation, and he had been out of the clinical sphere for the past six years doing research. He struggled to remember his clinical medicine and regain the basic hospital skills acquired in the first two years. He felt generally insecure on the wards.

One day George chose to follow a patient of one of the rotation directors, assuming he would be interested in teaching, a rarity among busy senior physicians. "I was scrubbed in for the delivery, and as the director stitched the episiotomy, he pimped me on all the muscles of the perineum, their innervation, and their blood supply. Of course I didn't remember. Then he said to me, 'How can you expect to be in the room with patients, scrubbed in on the delivery, when you don't know the anatomy? How can you expect to put in stitches if you don't even know where the nerves are running?' Then he let me tie a knot, and I got confused and started to tie a right-handed knot when I needed a left-handed knot, and he yelled at me again. He said, 'Obviously you haven't been practicing tying knots like I told you! You don't deserve to be in the room with this patient!' "

George was devastated by the comment. "Fortunately I have enough sense of self not to be completely destroyed by this. But the whole thing has just soured me on ob. I have spent the last few days just hiding out and reading. It's not like anyone ever taught us how to tie a knot. I learned out of a book and practiced on thread at home, but it didn't have the same texture as the suture. And no one ever lets us tie knots. If we don't ever get to do it, how are we supposed to get good at it? Anyway, with him yelling at me like that, there was no way I was going to do anything but choke."

Eric, a visiting fourth year doing an advanced ob rotation, had a similar experience. He was doing what was known as an interview rotation. Planning to specialize in ob-gyn, he hoped to come to Boston for his residency and used this rotation as an opportunity to learn about the hospital and let the residency directors get to know him.

"I was just in this delivery, and it was a total disaster," he told George and me after his first delivery. "I didn't even know I was supposed to pick up this patient because there was some confusion about which service she was on. So I just went in right for the delivery. The attending physician was already there, and he told me to scrub and do the delivery myself. But the last time I even saw a vaginal delivery was nine months ago during my rotation, so I didn't feel comfortable doing it, and he got really upset. He sent me to scrub, and then I didn't have a cap. So he yelled at me to go get a cap, but I couldn't find one. So I had to leave the room to try to find one, and by the time I came back, the patient delivered while I was still washing my hands. Afterward he told me, 'I'm never scrubbing on a case with you again.' It was such a disaster. I can't believe it. But I just didn't feel comfortable doing that delivery. I mean, it's been over nine months."

We looked at the board where all the patients and their physicians were listed to figure out which attending had yelled at him. "Wait, were you in room eight?" George asked him. "That was the chairman of the department!"

"Oh, my God. I blew it. I can't believe it. I blew it."

Negotiating the hospital could be tough for medical students. In most rotations we contributed little to patient care because we lacked the basic skills. Instead we presented an obstacle to efficiency. Senior physicians and residents, pressed for time, sometimes resented the teaching responsibility we represented. Far away from their medical school days, they often forgot how little they had known in their first clinical year. We struggled through short rotations and scrambled to learn the important issues, to absorb the essential questions, tests, and values, and to decipher the expectations. By the time we finally mastered enough information to feel comfortable and confident in the rotation, it was time to move on to another, completely unfamiliar domain of medicine.

At Patient-Doctor one afternoon, Alyssa offered a suggestion from an emergency department senior physician on how to deal with being yelled at. "It's like driving in Boston. You're sitting in traffic, and the cars behind you are honking their horns. They're honking and honking and honking for no good reason. It's their communal expression of the deep frustration and latent anger welling up inside. You know you're not doing anything wrong. You just listen to the honking and let it slide right by. It's just like driving in Boston. If someone slams into you, pretend you're in traffic with the honking horns," she advised.

Right, I thought. The only difference was, when I sat in traffic, I knew I understood the rules of the road just as well as the person honking behind me. But on the wards I didn't know anything about anything. And being yelled at was the validation of my worst fear: I didn't know enough to function in the clinical world.

One of my classmates was fed up with her chief resident, who made a habit of terrorizing the medical students and interns on the team. Leslie had a strong personality, with a loud voice and a boisterous laugh. She was very funny and enjoyed a good laugh, but if she was upset, she was not one to let it slide. One morning on rounds the chief resident asked a question about something she had been saying a few minutes earlier. Leslie didn't know the answer, sparking the chief resident's ire. "That's what I hate more than anything, when people don't listen to me! It drives me up a wall! There is no excuse for you not to listen!"

Leslie turned to her and said, "I was listening," she said. "I just didn't think it was important to remember." Afterward Leslie told me, "I saw my grade go up in smoke. But at that moment I just didn't care. I was so angry."

Paul, one of the chief residents on the gyn service, made me feel stupid. It became a self-fulfilling prophecy. Around him I said only stupid things. He could ask me the same question I had answered over and over again, and the minute he asked me, my mind went blank. I knew it was partly my fault. I had an unfortunately thin skin and took any criticism very much to heart. His opinion mattered too much.

Paul was a large Asian man. He wore trendy small oval wire-frame glasses, although I never saw him dressed in anything other than crumpled blue scrubs and broken plastic clogs wrapped with masking tape. He wasn't particularly brusque or short with me, but he had a way of brushing me off that made me feel belittled.

During my second afternoon on the service I bumped into Paul in the hallway. When he told me he was on his way to his afternoon clinic, I asked if I could join him.

"Well," he said, "I prefer to have some advance notice."

"If it's not convenient today, I'll come another day," I said.

"No, I guess it would be okay. Usually I like at least twenty-four hours' notice, but I can make an exception today." He started to walk away, and assuming that he was on his way to the clinic, I followed. I didn't know where the clinics were held. After I followed him down the hall and up a flight of stairs, he turned to me and asked, "Why are you following me around, Ellen?"

"I thought you were going to clinic," I stammered.

"Not yet." And he sped up to leave me behind.

One morning I was in the OR with Paul and a senior physician. I watched as they removed the patient's diseased ovary. As Paul and the senior physician dissected the rectus abdominis muscles, I knew he was going to pimp me.

"So, Ellen, can you tell me what this muscle is that I am dissecting here?" Paul asked me. I knew the muscle well because it was a favorite question for medical students in the OR. But the second Paul asked me, my mind went completely blank.

"I know what it is, and I know that it begins with *P*, but I just can't think of the name right now," I said.

"Think Egypt," the senior physician hinted.

That only confused me more. What was in Egypt? What relationship could there possibly be between the abdominal muscles and Egypt? My mind raced.

"I just can't tell you right now. I know it, but I just can't give you the name," I finally said.

"Pyramidalis," Paul said. "They are the pyramidalis muscles."

Of course.

In the hospitals my classmates and I were constantly questioned on our grasp of clinical medicine and pathological principles. Sometimes we were quizzed in the context of our patients, whose medical issues we were expected to understand, but more often the questioning was random. It was impossible to prepare for. This pimping, while not always the most difficult aspect of our existence in the wards, could be nerve-racking. But I quickly grew accustomed to the questioning and lost any sense of shame I might have possessed before entering the clinical years. I accumulated lots of experience saying, "I don't know," and gradually learned to forgive myself the gaps in my knowledge. Yet despite my pretense at having a good attitude, there were times when I could not easily dismiss my failures. Sometimes I thought I should have known the answer, or I was frustrated because I had known the information previously but had subsequently forgotten. Even worse, sometimes I actually did know the right answer but was so concerned about saying the wrong thing that I chose "I don't know" instead.

Carlos was pimped on anatomy during surgery, but because it was his first day on the neurosurgery service, he was unfamiliar with the basic surgeries. He could only reply, "I don't know."

"What's the first ligament I encounter in the back?"

"I don't know."

"What ligament is this?"

"I don't know?"

"Yes, you do. It's yellow."

"No, I still don't know."

"The ligamentum flavum."

Finally the resident standing next to him whispered, "You shouldn't say, 'I don't know.' You should guess *something* even if you know it's wrong."

In a later surgery another resident took pity on Carlos. As the surgeon asked questions, the resident stood behind him and whispered the answers.

While coming up with the right answer on demand could be difficult, by far the greater challenge on rotations was figuring out the expectations. There was always a defined but generally implicit list of requirements for us. The good medical student was able to decipher this unwritten list, perform the expected duties, and accumulate the required information with a minimum of reminders and reprimands.

Renu had started her surgery rotation at the same time I started my ob-gyn rotation. Classmates who had already finished their surgical rotations at her hospital uniformly hated the program, finding the surgical residents unpleasant to work with. But even with that reassurance, Renu continued to struggle in her rotation. Week after week in Patient-Doctor small-group sessions, she tearfully recounted yet another horrible experience.

"I was sick the first day and missed orientation, so I just showed up the second day and went to surgery," Renu told us early in October. "I didn't even know what kind of surgery I was going to be seeing, so I couldn't prepare. I really liked being in the operating room. I was standing next to the surgeon, and I was interested, so I kept asking, 'Oh, is this the vena cava?' and 'What are you cutting there?' and things like that. The surgeon didn't seem to mind. Afterward, the chief resident, who was also in the surgery, came up to me and said, 'Next time, if you can't think of anything intelligent to ask, just don't say anything at all.' I was so upset. I mean, how was I supposed to even know what surgery I'd be watching?"

"Well, you just have to understand that surgeons expect you to be silent," Alyssa said. "I was always silent. In my whole rotation I might have asked one question during a procedure. Even if the surgeons are talking about their golf game last week or something that sounds superfluous, you might not realize it, but they are really concentrating, and your questions might be distracting. And the surgeons, they expect you to know absolutely everything about the patient and the procedure before you walk in the door. They expect you to be able to write the patient's entire history up on the board if they asked. If you don't know the patient and the procedure in detail, then you don't deserve to be in the operating room."

Renu certainly had a hard time negotiating the operating room hierarchy, but Alyssa seemed to have taken the behavioral code a little too seriously. I cringed as Alyssa tried to help Renu understand how to act to get through her rotation, but it was easy to see how both could be a little misguided. At the beginning of each rotation the course director met with us and listed the expectations. But these expectations involved what topics we were expected to cover, what procedures they wanted us to learn, what percentage of our grade would be constituted by the final exam. No one ever told us how to behave: not our course directors at the beginning, or the residents and physicians on our teams, or our classmates ahead of us. And no matter how hard we tried, we all got it wrong at least some of the time.

Renu grew increasingly miserable as the weeks of her surgical rotation passed. She began to question seriously her choice to go into medicine. In the second to last week of her rotation, she came to Patient-Doctor only to tell us she was leaving. She had had a terrible weekend, she told us. She became so depressed that her husband, worried and at his wits' end, brought her to the hospital. She stayed in the hospital over the weekend, and she had decided to take a medical leave of absence. Would she come back? we asked.

Renu looked at us, her deep brown eyes glistening with tears. "I don't know."

Ironically, often it was not the clinical staff but our classmates in the rotation who made the experience so difficult. Grades mattered, especially for rotations in specialties that we considered for residency. High honors was a relative score, and we had to perform consistently above our peers to earn the highest grade. During surgery the course outline at one of the hospital sites stated that students were expected to take call every fourth night. However, none of the administrative staff ever specifically mentioned the call schedule and left it up to the students to figure out.

Alyssa was the first person in the three-month rotation to serve on surgery ward service. Rather than the suggested every-fourth-night call schedule for her ward month, Alyssa chose a grueling every-other-night call schedule. Every time she finished a thirty-six-hour day, she went home after sign-out rounds, slept the night, and returned the following morning for another thirty-six hours, weekends included. When Alyssa finished her month on the service, Jerry, the next of our classmates to work on the service, felt compelled to continue her pattern of every-other-night call. "Otherwise they'll think I'm a complete blowoff," he said.

Jerry was an older student with young children and a pregnant wife. Carlos bumped into Jerry often in the call room in the middle of the night when nothing was happening with the patients. But Jerry sat there night after night just to observe the every-other-night call schedule and appear as enthusiastic and eager as Alyssa.

Jorge, the third person to follow Jerry and Alyssa on the ward service, viewed the situation with trepidation. "Oh, God, I'm screwed."

He too felt compelled to take call every other night. He lasted about two call nights of every-other and then, fed up and exhausted, took call only every third night.

In the emergency department, where students were expected to follow a twenty-four-hours-on, twenty-four-hours-off schedule for three weeks, Alyssa also intensified her hours. During her twenty-four hours off, she headed off to the operating room to scrub in on extra cases.

Several classmates doing a neurology rotation, one of the lighter rotations, complained about a fourth-year classmate who took call every third night on a rotation where students were generally not expected to take overnight call at all. They were frustrated by his zealousness because they worried the clinical staff would consider them slackers in comparison or, even worse, would expect them to take on call responsibilities as well.

"You know, I just don't care," Melissa said to me. "I don't care what they think of me because there is no way I want to take call. I'm just going to go in, evaluate the patients I have to, and go home when I want to."

Later on they learned that the classmate taking the additional call responsibilities was Alyssa's husband.

"Why am I not surprised?" Melissa asked me.

*   *   *

The sheer bulk of hours spent in the hospital could be ominous. Once you were in the hospital, it was easy to forget that another world existed beyond. Caught up in the details of my patients, I found the hours passed quickly. But every once in a while the realization that I had been in the hospital since 5:00 in the morning and wouldn't be home until after 7:00 P.M. the following day overwhelmed me.

Despite the long hours, I actually enjoyed the night on call. I liked the rhythm of the hospital at night. The corridors were dimmed, and the visitors had gone. The senior physicians and house staff not on call were at home catching a few hours' rest before they returned early the next morning. A quiet calm blanketed the hospital. There was less time pressure on the interns and residents, so I enjoyed the extra time for teaching.

The most difficult time for me in the hospital was waking up in the morning after my call night. I woke up in an unfamiliar bed in an unfamiliar place. I had been wearing the same clothes for twenty-four hours. I hadn't gotten enough sleep, and I hadn't been outside the hospital since the previous morning. I knew I had a long day ahead. I struggled to prepare for rounds, feeling depressed. But once I was involved in my patients and rounds, my day picked up. By the afternoon I already anticipated coming home.

Weekend call was the worst. On Saturday and Sunday, no longer subject to the routine of the weekdays, I was overwhelmed by the bulk of unstructured hours that lay before me. My worst experience in the hospital was a Saturday-to-Sunday call on the gyn service. I spent twenty-four hours in the hospital and, in all that time, saw a total of two patients. By lunchtime I was on the verge of tears. Fortunately Carlos met me for lunch at the hospital cafeteria to cheer me up. I spent our two hours together crying into my bag of animal crackers in the cafeteria. All I wanted was to go home.

Pediatrics

I loved the children. I loved their problems. I loved their issues. I loved their parents. This is not to say that I liked every patient or every parent, only that I loved taking care of children. I didn't expect to like pediatrics as much as I did. While I had always enjoyed children, I had never thought about what it would be like to deal with them in a medical setting.

I did my pediatrics rotation in a tertiary care center, equipped with an intensive care unit and facilities to care for severely injured trauma victims. At this hospital many of the children had serious chronic diseases with complex medical issues. I saw children devastated by neurologic injury at birth, the mental retardation—cerebral palsy ("MR-CP") kids. They were contorted by muscle contractions from the cerebral palsy and unaware of their surroundings because of the severe mental retardation. Claire was deformed, malnourished, and retarded from congenital cytomegalovirus infection. Julie, a bright and capable fourteen-year-old, was weakened by muscular dystrophy. She drew from her personal experiences to help her seven-year-old roommate with cystic fibrosis cope with nebulizer treatments. "Even though we have different diseases, I understand what you're going through. I have to do the same treatments for my lungs too," Julie told her.

We also saw children who had more severe forms of common illnesses, such as asthma and pneumonia. In the "Empyema Room," Evan and Dimitri both struggled to recover from their pneumonias complicated by infected fluid that collected in the lining of their lungs, compressing and collapsing them. Alexis, an eighteen-month-old toddler, suffered from a particularly severe and lengthy bout of the croup.

We saw children whom we couldn't diagnose. There was two-year-old Ibrahim with cyclical episodes during which he turned red, then blue as he stopped breathing, vomited, and returned to normal. There was Matt with his leg pain, and Katy with her stomach pain, both most likely psychiatric in origin.

The pediatric floors of the hospital were divided into pods, each identified by its own slightly different wallpaper runner. Mine had cartoon jungle animals running along a wide green stripe. Each pod consisted of a set of rooms arranged in an L around a central nurses' station. Toys often cluttered the wide hallway. The younger children particularly loved a yellow plastic car they sat in and propelled by walking with their feet. We often had traffic jams during rounds as the yellow car coming in one direction met a stroller traveling in the opposite direction while ten people tried to pass by. IV poles were another favorite. Each metal pole had a tripod base on wheels with a horizontal circular bar connecting the three legs. The older kids loved to take a running start and then stand on the horizontal bar as the pole coasted down the hall. Although the nurses often gave out "speeding" tickets and tickets for "reckless driving," their efforts were largely in vain.

Each pod had its own playroom. Ours was a bright corner room with lots of windows. A Christmas tree had been set up in the far corner with multicolored shiny balls and plenty of tinsel. It wasn't as cute as the tree in the pod next door, decorated in a *101 Dalmatians* theme, draped with stuffed dalmatians and dog biscuits, but it lent the room a festive cheer. The room had four round tables with child-size chairs. An arts and crafts project was always set up on the far table, and board games were stacked on shelves along the sides of the room. Milk crates along the walls were stuffed with dolls. A computer in the far corner of the room was set up with children's games and educational software. Children posing an infection risk to other patients were barred from the playroom, and it was torture for them to be stuck in their rooms all day. Each room had a TV and VCR to help offset boredom. *Toy Story* was the overwhelming favorite movie. One poor mother complained bitterly by the third day of the continuous repeats her three-year-old son insisted on.

The most feared spot on the pod was the treatment room. The small room had an examining table in the center, two small carts with blue drawers, and a counter on the far wall. An orange plastic briefcase decorated with stickers sat on the counter. The treatment room looked innocuous to the unsuspecting eye. But hidden inside the cabinets and the drawers was every manner of tiny IV needles, syringes, and blood-drawing equipment. While blood draws and IV placements were done at the bedside without a second thought on adult wards, these simple procedures frightened the children. They couldn't understand why they needed them, and with their undeveloped concept of time, they didn't realize the discomfort would be temporary. Procedures were never done at the bedside to preserve a safe haven for the children; no one would hurt them when they were in bed. Instead their parents brought them to the treatment room, the orange briefcase was opened, and while the parents tried to comfort their children and distract them with the toys and stickers in the briefcase, we attempted the blood draws and IV placements.

As much as I loved the kids, I hated the procedures, especially in the middle of the night. The children screamed as we struggled to find their tiny veins. I hated the tortured looks on their parents' faces as they attempted to calm their children. One angry, hurting father grabbed the IV out of the intern's hand just as she got it into the vein, screaming, "No more of this! I don't care. No more! I'll take full responsibility. I'll take full responsibility for him missing his antibiotics tonight!" and flung the bloodied IV catheter across the treatment room.

The children were often terrified of the white coat, and most of the residents and physicians chose not to wear them. I was relieved to put aside my coat with its uncomfortable power connotations. As much as I looked to my coat to provide a sense of community and legitimacy in a world where I felt I didn't quite belong, I grew to dislike the formality it brought to the patient-doctor relationship. I felt it made me a little less human and a little more generic. Once I established a relationship with a patient, the significance of the white coat faded into the background. But I preferred a more casual style. Without white coats, their length distinctions, and the connotations of stuffed pockets, it also became more difficult to recognize at a glance where people ranked in the hierarchy. Ironically, I felt more integrated into the medical community without my white coat.

Caring for children introduced nuances into the patient-doctor relationship. The medical issues themselves gained intensity when a child's life was at stake, and it was all that much more important to preserve both life and quality of life in our pediatric patients. The clinical relationship was established with both the parent and the child. But as the children grew older and began to assume more responsibility for their lives, it was sometimes hard to incorporate this new independence and autonomy into the tripartite medical relationship. This was especially difficult when caring for the many HIV and AIDS patients we had in our pod. Several of the parents wanted to protect their children from their diagnoses, yet we realized that these children were likely aware of their illnesses. As medical caregivers the residents, physicians, and I were caught between parent and child. We had to honor the parent's prerogative and were complicit in the silence that reinforced the shame of the child's situation and prohibited her from discussing her fears openly.

*   *   *

Despite horrible disease, the kids found ways to preserve their childishness. Samantha and Alexa, eight-year-old identical twins with severe cystic fibrosis, spent much of their year in the hospital. Cystic fibrosis, by inhibiting normal secretions in many areas of the body, had left their lungs prone to infection and prevented their bodies from absorbing food normally. Class pictures and annual holiday photos of the two girls dating back to their toddler years were posted on one of the bulletin boards in the playroom. They had a list written on orange construction paper posted on their door. Headed "Things That Are Good to Do in the Hospital," the list included: "Be nice to my sister. Cooperate with Chest PT [physical therapy]. Eat my whole dinner." Just below it hung a sign that said: "Things Not to Do in the Hospital: Hit my sister. Play jump rope with my IV. Ride on my IV pole."

I was amazed by how much I enjoyed taking care of children. For kids, life didn't stop when they entered the hospital. They still appreciated the pleasure of a color, the comfort of a familiar toy, the first snowfall of the season. This fascination helped me remember to celebrate life as well. I loved that birthday parties were important and that the residents and physicians worked hard to get kids home in time for them. I loved that we could take a few minutes to listen to a joke or play a game with a child on rounds.

I also enjoyed the family focus in pediatrics; the parents and siblings were as important to each encounter as the patients themselves. I found myself particularly fascinated by the parent-child relationship under the stress of chronic illness. This recognition of family had been so important to my hospice experience, and I began to realize how much I'd missed this aspect of patient care during my first months of rotations.

I felt happy during pediatrics. Until that rotation I had been absolutely certain I would choose internal medicine to pursue a career in hospice and palliative care. But for the first time I wasn't so sure.

Jamie

•

I could not get Jamie out of my mind. His image replayed in my head whenever I let my mind wander. At night I dreamed I sat with him in his white hospital bed and read him stories. I dreamed I rushed him to the hospital but lost my way. Jamie was hidden in the backseat; only his IV pole with a small hand clutching it jutted up above the high back of the front seat, reminding me of the urgency of my mission.

Jamie had Caroli syndrome. An autosomal recessive disorder, it caused severe liver and kidney disease. At six and a half years old, he was in end-stage liver and kidney failure awaiting a transplant. His liver, stiffened by the disease, was unable to accommodate all the blood as it passed through for detoxification on its way back to his heart. Instead much of Jamie's blood bypassed the liver and returned to the heart via the vessels of the esophagus. The day before he arrived in Boston, these vessels, engorged with the extra blood, ruptured and bled. This complication of liver disease is often fatal. A person can lose half his blood volume before even reaching the hospital. After blood transfusions in a community hospital, Jamie was flown to Boston by helicopter for sclerotherapy, a special procedure using chemical irritants to promote scar formation and reinforce engorged vessels so they would no longer bleed. This was not a new experience for Jamie; he had been MedFlighted to Boston once before for this procedure. That stay had been complicated by a serious blood infection, and he had been hospitalized for almost two months. Over the last year he had also had several smaller episodes of bleeding.

Karen, the senior resident, suggested that I take on Jamie as my patient. I was nervous. How could I manage a patient who might bleed at any minute? But afraid to refuse and definitely not one to back away from a challenge, I said nothing.

Jamie arrived in Boston just after dinnertime. Karen, Danielle, who was the intern, and I went to meet and examine Jamie. The liver specialist was already there. Technically, because Jamie was my admission, I should have directed the interview and exam. But I hesitated when we met him, worried I would prolong the admission and certain I would annoy my senior. I was afraid to talk to Jamie because I might be superfluous, and afraid to touch him because it took me longer than the others to identify structures and the like.

Because his mother was still en route to the hospital, Jamie was by himself. He had been placed in a big room with four beds. He lay in the far corner. The room was dimmed except for the small corner of light from his area. I expected to see a sick, frightened child lying in bed, but Jamie reclined on three pillows, his legs bent up and crossed at the knees, with the dangling foot rhythmically kicking. A large teddy bear, nearly as big as Jamie himself, with a big red bow at the neck, lay in bed next to him. Jamie chatted easily with Mary, his nurse, and directed her on how to orient the various tubes emanating from his tiny body. He was clearly in control of the situation.

Although six and a half, Jamie was only the size of a four-year-old. As we came closer, I saw that his skinny little arms and legs were covered with a thick layer of dark hair. His protuberant belly swelled incongruously beneath his miniature yellow hospital gown festooned with romping clowns. He wore a diaper. A thick shag of brown hair and bushy eyebrows dominated his large head, which made his small, sharp nose and pointy ears look even smaller by comparison. One of his medications to control high blood pressure, minoxidil, the same active ingredient as in the hair replacement treatment Rogaine, had caused the hairiness. His dark eyes regarded us skeptically as he took us in. His lips were crusted with cracked maroon blood. Jamie's odd proportions and hairy body made him look like an old elf. I felt as though he could bewitch us at any moment.

Jamie came to us with two IVs, one in each arm. We wanted to make sure that both IVs worked appropriately because they were essential if he had another bleed and needed immediate fluid replacement and blood transfusion. As we played with Jamie and his teddy bear, Mary went to flush saline through one of the IVs to make sure it was still running. She depressed the plunger on the syringe connected to one of the lines, but the fluid did not budge. An open, well-functioning catheter should easily accommodate the small amount of fluid in the syringe. Mary looked up at us and shook her head. She pushed one more time with no result. Jamie momentarily stopped playing with the bear and looked over at Mary. "It won't flush?" Jamie asked.

"No, hon, I can't get it to flush," she said.

"So the IV has to come out?" Jamie asked.

"Yes, I think so," Mary said.

"But you'll give me another one, right?" he asked.

"Oh, yes," said the senior. "Don't you worry. We'll get you another one."

"Okay. That's good."

"You know what, Jamie? You're such a special boy that I have a present for you," Karen said.

"A present? Where is it?" Jamie was interested.

"I have to go get it. It's in my locker. But I think you'll like it."

Karen returned in a few moments with a light stick, one of the various toys and trinkets she stored in her locker to give out. She made a tear in the package so that Jamie could open it. We all reached in with our stethoscopes while Jamie fumbled with the packaging.

"Are you having trouble with that? Here, why don't I open it a little more?" Karen broadened the small tear and gave the stick back to Jamie.

We reached in to feel his protuberant abdomen. He had a long, thin scar indicating where his left kidney had once been. The delicate scar reached around his back and strained for his belly button. The edge of his liver, normally felt just below the last rib, extended all the way down to his pelvis.

Jamie still couldn't open the light stick, so I took the package from him and ripped off the top. He took it back and pulled out the eight-inch plastic rod filled with electric orange fluid. He tipped the rod back and forth, watching the orange fluid sway within its plastic confines.

"Look," Karen said, "if you bend it, it will light up. Like this." She showed him how to place his hands and bend the tube. But he could not bend it hard enough.

"Here, how about if you move your hands closer together? I'll help you." With the two of them bending together, the rod let out a quick crack and the electric orange fluid suddenly fluoresced purple. Jamie's fingers glowed red around the tube of light.

Jamie's mouth opened in amazement, forming a crusty maroon O. "Wow!" He held up the rod and moved it around in the air, and the rod left a faint trail of purple in the dim air. Then he put the rod back on the bed and tried to bend it again.

"What are you doing, Jamie?" Karen asked. "You don't want to break it."

"I want another color!"

"You can only do that once," Karen said. "But here, let's see green." She put the fluorescent wand under the edge of her shirt. Jamie was thrilled. He took the rod back and stuck it underneath his hospital gown. "Blue!" he exclaimed triumphantly.

"All right, Jamie. Unless you have any questions, we're all finished with you for now. Is your mommy coming?"

"She'll be here later," he told us. As we left his corner of the room, he waved good-bye to us and began chatting with Mary again.

Jamie's mother didn't arrive until late in the night. I went in to see her twice, but she was on the phone. Jamie lay quietly in bed, watching *Toy Story*. It was after 11:00 P.M. when she finally settled Jamie in for bed. His corner of the room was dim now except for a small light behind the bed. He lay on his side in the center of his bed, still awake, sucking his first two fingers. His mother adjusted the plastic line attached to his G-tube, providing permanent access directly into the stomach (gastrium) for feeding purposes. She pinned the clear plastic line to the sheet so it wouldn't pull on him during the night. "You sleep in bed with me, please," Jamie said.

"There's not enough room for you and the bear *and* me!" his mother said.

"Yes, there is! You sleep here! Please."

"Okay, but you're going to have to move over if you expect me to fit in here! Which side do you want?" His mother unpinned the G-tube line. "Move over, Jamie."

"Uuh!" Jamie said.

"James, please move over so I can fit in."

"UuhUuhUH!"

"James, I asked nicely. Don't uuh me! This is not acceptable behavior." Jamie finally obliged and moved a little closer toward his teddy bear. His mom crossed to the other side of the bed and repinned the G-tube line. Once it was adjusted, Jamie quickly drifted off to sleep, and his mother came over to talk to me.

She was a small woman, and her large blue eyes were a stark contrast with Jamie's small dark eyes. Her face glowed pale in the dim light. Her thin, wavy brown hair fell below her shoulders.

She wore a black T-shirt with the oversize face of a white cat printed on it, tucked into slightly tight gray jeans, and black high-top Reebok sneakers.

"I know it's late, and you've had a long day," I said, "but I was hoping you could tell me a little bit about what happened to Jamie that brought him to the hospital."

"We went to the doctor's in the morning. Jamie had a 'crit of twenty-eight point six, which was just a little lower than his usual in the thirties," his mother said. "But he was doing just fine, and the doctor gave him a clean bill of health. So we went out in the afternoon to the mall. We came home at about five-thirty. Just after we got home, Jamie began complaining that he was tired, so I told him to go to his room and put on his pajamas. I told him I'd be up in a few minutes to put him in bed. But then it was real quiet upstairs, and I went in to check on him because it was so quiet, you know? When I got to his room, he wasn't there. I found him in the bathroom, still dressed. He hadn't even changed into his pajamas. He was leaning over the toilet, and he said he was feeling like he had to throw up, but nothing would come up. And then, all of a sudden, he vomited up bright red blood and clots right into the toilet. I picked him up to take him to the hospital, and he was limp and pale as a ghost. We ended up in the ICU. They gave him a transfusion there. And now here we are."

"How was Jamie diagnosed with his disorder?"

"At birth. Actually, before birth, on the ultrasound. They could see the cysts in his kidneys."

"And you don't have any family history of this?"

"No, I have one kidney, but that's it. Can't you get this out of his chart? I don't mean to be rude and all, but I've had a long day, and I'm real tired."

"I know you've had a tough day, and I don't mean to bother you. I'll be quick, I promise." I still needed to get some information about their medical history and living situation. But she and I both were tired. It could probably wait.

"You know, I was so impressed with Jamie when we met him

earlier this evening, before you got here," I said instead. "He knew the whole routine. He was so confident and at ease, talking to all of us and answering all the questions. Most kids would have been anxious and afraid without their parents, but he was totally on top of the situation. He's so bright."

Her blue eyes held the hint of a smile, and her face softened. "He was? I was so worried about sending him in that helicopter. I promised myself that this time I wouldn't cry. I promised myself. But the minute they put him on that gurney and he started crying, I just couldn't help it. I was bawling. I was frantic driving up to Boston. Then I get here, and he's happily playing away. I asked him how his trip was, and he said, 'Oh, I had a lot of fun. The man in the helicopter, he started singing Christmas songs, and then I started singing, and we sang songs the whole way to Boston.' And I was like, here I am rushing to Boston because I am worried that he'll get here and be all alone, and the whole time he's having a blast singing Christmas carols!"

His mother continued to talk in earnest, wandering from story to story. "You know, we've always been honest with Jamie. We don't hide anything from him. Today, in the hospital, he turned to his nana, and he said, 'Nana, I'm dying.' And she said to him, 'No, don't say that, Jamie. You're going to outlive me.' So he turned to her and said, 'But, Nana, you're an old lady.' Then, later on, when he found out that he was going on the helicopter, he turned to me and said, 'Mommy, I'm going on the helicopter, right?' And I said, 'Yes.' And he knew. He said to me, 'That's for people who are real bad off, right?' And I said, 'Yes.' So he said to me, 'That's for people who are dying, right?' I said, 'Well, sometimes.' And he said, 'Mommy, am I dying now?' I told him, 'Oh, no, you're not dying now. They just want to get you to Boston quickly so they can make sure that you're okay.'

"Today in the hospital, after he found out he was coming to Boston, some of the doctors came by to see him. He told them, 'I know that Mary, my nurse at Boston, loves me.' And they said, 'Oh.' Then he said, 'You want to know how I know she loves me?' So they said, 'How?' So Jamie said, 'Because last time when

I was there and I was so sick that I couldn't even open my eyes and all I could do was cry, Mary sat and held me until they took me to the ICU.' And I said, 'But, Jamie, we all held you. You just don't remember.' We did; we all held him and passed him back and forth between us."

"This Mary? The Mary who is on tonight?"

"Yes, that's her. Before he got on the helicopter, he asked me to buy a flower for her, but I couldn't get one tonight. He hasn't asked for it yet, so I think maybe he forgot."

Just before midnight I finally interrupted his mom, anxious to let her get some sleep and eager to write my note and be done for the day. During the course of our conversation I learned that Jamie's parents were divorced. He lived with his mother and her fiancé; two older sisters lived with another relative. His mother had custody of the girls on the weekends and had been planning a birthday party for the eleven-year-old on Sunday. When she called her daughter that evening, her daughter informed her that she had better get those doctors to let Jamie go home by the weekend.

As I left Jamie and his mother, I felt calm after my turbulent evening. I felt the profound sense of peace that comes after long, deep conversation. I felt connected.

The next day Jamie felt more himself. He had his sclerotherapy procedure scheduled for the late afternoon. In the middle of the morning I looked for Jamie and his mother to discuss the plan for the day. I found them in the playroom. Jamie's mother had laid him in a wooden hospital bed/wagon, and she pulled his IV pole along with them. His teddy bear was tucked in the wagon beside him. Several of the interns and one of the team physicians were grouped around him in the playroom. He had just finished tracing a Christmas tree in green marker on a piece of orange construction paper. "Jamie, what's your bear's name?" one of the interns asked him.

Jamie picked up the bear and studied its ID bracelet buried in the fur of the right paw. "It's the same name, James! It's the same name, James! It's . . ."

"The nurse gave the bear a matching ID bracelet so that today when he goes to his procedure, they won't make a mistake and give the bear back to the wrong Jamie," his mother explained.

I didn't see Jamie for the rest of the day. When I left for home, he was still in the operating room. When I arrived the following morning to preround, I learned the sclerotherapy procedure had gone well. He had even been fed the night before: mashed potatoes. I was excited. I hoped he would make it home in time for his sister's birthday party after all.

During walk rounds we knocked on his door to say good morning. "Hello, Jamie. Good morning!" I called from the doorway. "We're all here to say hello."

"No! Go away, please," he said.

"How about if just two of us come in? I heard that you got to eat mashed potatoes yesterday," I said.

"Yeah! I get to eat again today!"

"Are mashed potatoes your favorite?"

"Mashed potatoooes, chicken wiiings, and graavy, and mashed potatoooes, and chicken wiiings . . ."

"Well, Jamie, it sounds like you are doing much better."

"No, wait, I'm not finished! Mashed potatooes, and graavy—"

"Okay, we'll have to talk some more later."

In the midafternoon I went to examine Jamie and found him asleep in his bed. His mother was sitting, watching the video channel VH1 while he slept. "Sometimes he likes to listen to music while he sleeps," she explained.

I knew if I was going to be a "good" medical student, I should see what the interns needed me to do to help out the team. But I wanted to talk to Jamie's mom, so I stayed anyway. I was curious to hear how Jamie dealt with school.

Although Jamie had enrolled in kindergarten the year before, his mother told me, he had missed so many days of school from being in the hospital that the school asked him to repeat. Finally, as he got sicker and more fatigued, she took him out of school and arranged for home tutoring instead. "He likes it a lot better this way. He just couldn't keep up with the kids anymore. But he does great at his schoolwork. He's doing first- and second-grade level work at home," his mother said.

I asked her if he had many friends from school. "Oh, yes," she said. "You know Jamie. He has lots of friends. A lot of the kids that go to his school live in our apartment building, and they come by to see him all the time."

"But it must have been difficult for him at school, being small and sick a lot of the time."

"Oh, yes. The kids there, they called him Werewolf and Wolf-boy. They used to come running up behind him and try to pull up his shirt so they could see his hairy back. Last year, about a week before his first bleed, I was coming to pick him up. Just as I got there, while I was watching, a sixth grader came running down the hill and drop-kicked Jamie in his back. Jamie didn't have time to put out his hands to break his fall, so he landed hard on his stomach, and he scraped his arm. I know that Jamie would have bled anyway, but in his mind it's linked. He says to me, 'That mean boy, he made me bleed.'"

"And going to school with diapers must have been hard for him too. He must have realized that the other children didn't wear diapers."

"Oh, Jamie doesn't usually wear diapers. This is new. He only started wearing them when he sleeps since he got real sick. He just can't wake himself up in the night anymore. Now he wears a diaper at night, and I put a pull-up on him before his nap. He hates the diaper. He complains. And I tell him, 'But, Jamie, this is for medical reasons.' And he says, 'Mom, don't you think I know that?'"

By dinnertime on Friday we were able to stop Jamie's extra medications. After an additional twelve hours of observation, he was discharged on Saturday morning. He made it home in time for his sister's birthday party.

AIDS

During my month on the pediatric service, there was always at least one HIV-positive patient and often two or three. All the patients I saw had acquired HIV at birth. There was a wide range of disease; some of the children were sick and dying while others were only mildly affected as yet. But with the new protease inhibitors, there was a refreshing sense of hope for some of these children.

Maria had a hard time taking her medicines. Only five years old, she still couldn't swallow pills. While she was very cooperative and did her best to swallow even the most foul-tasting potions, indinavir, one of the new protease inhibitors, was her nemesis.

"As soon as she can take her indinavir, she is good to go home," Karen told Amanda, her nurse, during work rounds in the morning.

"Poor thing, she's really trying, and she wants to please us, but it just tastes awful. I've tried mixing it in everything," her nurse said.

"How about chocolate syrup?" someone said.

"This medication is a hundred dollars a dose. At some point we're going to have to cut our losses. We can't keep on throwing this medicine out if we're not going to get it into her," said an intern working at the computer.

"You know, all the taste buds are at the front of the tongue. How about mixing it in Coke syrup and using a syringe to squirt it all the way in the back of her mouth, past her taste buds?" someone else said.

"That's brilliant. Okay, I know exactly how I'm going to do it." Amanda rushed out of the room.

Ten minutes later Amanda returned to the workroom carrying Maria on her hip. Maria had a big grin, and her smiling cheeks squished her bright brown eyes into a narrow squint. Her long brown hair hung halfway down her back. She hugged Amanda's neck. "Maria has something to tell you all," Amanda said.

"I took my medicine!" Maria beamed.

"You *what?*" the intern said.

"I took my medicine!"

"You good girl! I'm so impressed!"

The senior walked into the room.

"Guess what!" Maria said to Karen. "I took my medicine. Amanda was so proud of me!"

"We're all so proud of you! And you know, this means you get to go home today. Do you want to go home?"

"Yes," Maria said, and then buried her head in Amanda's neck.

Maria's grandmother came a few hours later to pick her up. A tall woman, she wore her graying sandy brown hair in an unstyled Farrah Fawcett cut. She wore an old navy blue three-button work shirt with black jeans and dirty work boots. "Are you ready to discharge her?" she asked.

"We have to work on her discharge papers, and we need to

straighten out her ganciclovir dosing before she goes home. So it will probably be at least an hour before we can actually get her out of here," Danielle, the intern, told her.

"Now, about her new medication, I don't know how I'm going to give it to her. She already has the AZT and D4T to take in the morning and evening, but what about this new one? Can I give all three at the same time? I can't be giving her extra medicine in the morning because I have to leave the house by six-thirty to get to work. And how about the noon dose? How am I supposed to give her that when I work all day? Jesus, I just don't know how I'm going to handle all this," her grandmother said.

"We'll make sure that you have a workable plan before you go home. Are you healthy now, or do you have AIDS too?" Danielle asked.

"What do you mean? I'm her grandmother. I'm healthy."

"Right. Of course there's no reason you would be sick. Is her mother still alive?"

"Her mother has done everything wrong. Crack, coke, heroin—you name it, she's done it. And believe it or not, she's still healthy. Hasn't been sick a day in her life. She's off God knows where. Maria has a twelve-year-old brother who has AIDS too. Maria lives with us, and he lives with his uncle. It's a mess."

Annie also had AIDS, but she didn't know it. Or at least no one had told her. Only ten years old, she was a "frequent flier" on the pediatric service. Although she had been healthy for much of her life, her list of medications rapidly expanded as her CD4 count plummeted. The CD4 count represented the number of a particular type of white blood cells that HIV infected and killed preferentially. The number of CD4 cells was inverse to the severity of disease and served as a rough guideline for the functional capabilities of the immune system. Annie's hospital admissions had steadily crept up over the past three years, and this was her second admission in three months. This time she was admitted for inflammation of the pancreas, causing severe abdominal pain and vomiting. One of the anti-HIV medications had probably caused this acute episode. A bacterial infection in her bowel caused bloody diarrhea and made her an infectious risk to the other children. She was confined to her room.

Annie lived with her mother and grandmother, and although her grandmother helped give Annie her medications, her mother thought it imperative that the grandmother not know the diagnosis. But now that her mother was getting sicker and spending more time in the hospital herself, Annie's grandmother assumed greater responsibility for both their care. Annie's primary care physician, Dr. Moreau, who coincidentally also happened to be the senior physician for the team, had tried for years to convince her mother to tell Annie.

"I would always get just so far in the conversation, and she would give me big signals that I had probed too far. She was never willing to accept any help," Dr. Moreau said.

Annie was a quiet, thin child with pale brown skin and deep black eyes. Her hair was always pulled neatly into two fluffy pigtails. She had a slight overbite and crooked white teeth, and her chronically inflamed salivary glands gave her face the appearance of chubbiness. She was always pleasant and cooperative when we came in for rounds, but we could never get her to talk.

"This is horrible. I can't take care of this child who doesn't know her diagnosis!" Danielle was appalled. "How could her mother not tell her? What is she thinking? What am I going to do if she asks me what is wrong with her? What is the policy on this anyway?"

"The general rule of thumb is that we tell parents we will do our best to honor their wishes, but we just can't promise. Mistakes happen, especially on a busy service with multiple caregivers. But what *are* you going to say if she asks you what's wrong with her?" Karen answered.

"Well, I guess that I would just tell her that one of her medications made her tummy sick, and we are going to help her get better," Danielle said.

But during her entire week in the hospital Annie never asked.

*   *   *

One night when I was on call, the nurse told Danielle and me that Annie had just had a large, bloody bowel movement and was terrified. She wanted one of the doctors to come in and tell her she was okay. Danielle and I walked into Annie's room to find her in her mother's lap. She tried her best to squeeze her long ten-year-old body into the small sanctuary of her mother's lap. She didn't want to look at us when we came in. Her mother was a small woman. Her shoulder-length straightened hair was carefully arranged and seemed to match exactly her dark skin. She looked at us with anxious eyes, the white of her eyes a stark contrast with the monochromatic brown of the rest of her face. Years of worry had sculpted her taut skin into pronounced ridges and valleys. We reassured Annie and then left the room. I never saw her mother again.

"Do you think Annie really doesn't know that she has AIDS?" Dr. Moreau asked one morning on student rounds when we discussed her case.

"I think she knows," I said. "If she doesn't know the name of her disease, which I think she might, then I think she realizes she is sick and dying."

"I agree," Dr. Moreau said. "I'm sure that somewhere along the way she figured out that she has AIDS. There have been studies on this in kids with cancer. Even when the parents and caregivers have done their best to withhold the diagnosis, people talk. The kids figure out what the other kids in the waiting room with them have, or they read their diagnosis off a chart, or they overhear a conversation. But since their parents have made it a big secret, they can't talk about it. They feel they have to play along and prevent their parents from realizing that they know. So they don't ask. There have been children on their deathbeds who have never been told, and they *still* don't ask. I think I have let this go on too long for Annie. This hospitalization is making me realize that I really have to push her mother to tell her. I need to explore further why it is so important to her that Annie not know. This just can't continue," he said.

"And her grandmother," my classmate added. "Her grandmother will presumably be her caregiver after her mother dies. She will need to know too."

I was frustrated with Annie's mother for handling the situation badly. I understood her mother's fear of the stigma associated with AIDS and her desire to protect Annie. Yet how could she rob her child of language? How could she rob her of an explanation? I knew that we as acute caregivers did not have sufficient depth in our relationships with either Annie or her mother to address this issue. I knew an acute hospitalization was not necessarily the best circumstance under which to confront Annie with this information. I recognized we did not have all the right answers. But for Annie we were powerless. We could only watch and hope that her mother would eventually make the right decision and release Annie from the rigid confines of her dark secret.

At five years of age Jonathan was dying. He had been diagnosed with HIV when he was two months old and developed pneumonia. At that time his mother told the doctors that she was HIV positive, leading them to the diagnosis of her son. When he was eight months old, Jonathan stopped developing normally. He never sat up. He never learned to grasp. He never learned to talk. Jonathan had HIV encephalopathy, a progressive deterioration of the brain. When he was two, he developed HIV cardiomyopathy; the HIV had weakened his heart. Now, at five, he was developmentally equivalent to a six-month-old and the size of a small three-year-old. His body was rigid from perpetually contracted muscles, and he lay stiff, stretched across his mother's lap, his red pacifier moving rhythmically up and down as he sucked. His most recent CD4 count four months before was 17; a CD4 less than 200 defined AIDS, and a CD4 count less than 50 identified extremely immunocompromised people at tremendous risk for developing infectious diseases. This time he was admitted for pneumonia.

"We went to New York to visit relatives last week, and everybody was sick with the flu. When we got home, he was doing fine until two days ago, when he developed a low-grade fever. I just gave him Tylenol, but today his nostrils were flaring, he was grunting with each breath, and I saw some retractions," his mother told us.

"Wow, you really know all the signs of respiratory distress," I said, impressed.

"Oh, yes. We've been through this many times with Jonathan," his mother replied.

"Been there, done that," chimed in his ten-year-old sister, lying on his bed watching a video of *Aladdin*. She was the only member of the family who did not have AIDS.

When Karen, Danielle, and I went to examine Jonathan, he lay outstretched across his mother's lap. He watched us carefully with his bright brown eyes. His thin brown curls were moist with sweat, and his light brown skin glistened. By the rapid movements of his chest, we saw he was breathing with difficulty. His neck muscles tightened with every breath, helping his respiratory muscles do their work. As we reached in with our stethoscopes, he raised his rigid arms in an effort to push us away.

"Oh, and by the way, he also has some thrush now in his mouth, so I think he needs a prescription for that flucon, flucan—"

"Fluconazole?"

"Yeah, that's it."

"Well, we'll have to look in his mouth first." Karen tried to use a tongue depressor to open his mouth, but as soon as Jonathan saw the wooden blade, he clamped his mouth shut.

"He's not going to let you look. You can trust me. He's a stubborn one," his mother said.

Karen tried one more time. "Would you mind if we pinch his nose? That's how we usually get the little guys to open up. Eventually he has to take a breath. I know it might sound cruel, but it's usually the best way," she said.

"Be my guest. You can try."

Karen pinched Jonathan's nose. His bright brown eyes crossed and seemed to stare at the bridge of his nose. He struggled not to take a breath. Then, when he could no longer hold out, he opened his mouth just the tiniest crack to let in a breath of air and clamped it tight again.

"Okay, it's clear that he's not going to let us do this tonight. We can try to look again tomorrow."

Jonathan, his sister, and his mother lived in transitional housing. Although his father was actively involved in parenting Jonathan and his sister, he did not live with the family. The day after Jonathan was admitted to our hospital, his father was hospitalized with *Pneumocystis carinii* pneumonia, a ubiquitous pathogen that causes disease only in the most severely immunocompromised patients.

"Are you clean now? Or are you still using drugs?" Karen asked Jonathan's mother.

"What do you mean? I don't do drugs! I never did none of that!"

"Oh, I'm sorry, I just thought that your housing program was for drug rehabilitation."

"No, it's for women who are looking for housing and need someplace to stay in the meantime," his mother said.

"Well, then how did you get AIDS?"

"I got it sexually, from my husband."

"Oh, my God. I'm so sorry. That's horrible." Karen's face softened, and for the first time during the interview, she leaned in toward her. Now, all of a sudden, this mother was transformed from reckless perpetrator to innocent victim.

I was uncomfortable with this conversation. I felt we had abused the power of the white coat and recklessly transgressed her personal boundary.

"Does your husband use drugs?" Karen continued.

"No, he got it sexually too."

"How horrible. I'm sorry. I'm so sorry."

"Yeah, well, we've learned to deal with it."

One of the issues to be resolved for Jonathan during this admission was his code status. He was currently full code, which meant that all invasive procedures and technologies would be used to prolong Jonathan's life. But Jonathan's condition was rapidly deteriorating, and it did not seem he had much longer to live.

Initially it seemed inconceivable to me that this child would be anything other than DNR (do not resuscitate). He couldn't speak; he couldn't sit up; he couldn't hold his mother's hand. And he was sick. What were we trying to save? I wondered. Of course I understood this was his mother's choice. I understood she might not be ready to let him go. Nonetheless I could not understand what kind of life she wanted to prolong.

As I watched Jonathan and his mother over their days in the hospital, I began to understand. The following afternoon I went into Jonathan's room to do his daily exam. I found his mother lying in the enormous hospital bed with Jonathan. She hadn't noticed me and was talking softly to Jonathan and stroking his head. Jonathan watched her with his enormous brown eyes, his red pacifier still bobbing up and down. Jonathan cooed as his mother kissed him on the forehead. "You're my boy, Jonathan, aren't you? You're my boy."

On another afternoon I came into their room to find her asleep, curled around Jonathan. She had left the TV on but without the sound. Jonathan was wide-awake, intently watching the rapidly changing images on the TV. When he noticed I had come into the room, he eyed me suspiciously. As I approached the bed, he whined, and his eyes crossed over the bridge of his nose. His mother woke with a start and began to stroke his forehead again. "What's the matter, baby? What's going on?"

On the day of his discharge I entered Jonathan's room to find him alone. His dark form rested in a cloud of white pillows. Alone, he was suddenly dwarfed in the vast expanse of bed. The TV played without the sound. As I entered the room, Jonathan turned to look at me. His bobbing pacifier beat even faster, and his eyes crossed. He began to complain.

"Hi, Jonathan. How are you today," I crooned. "Are you ready to go home today? Where's your mom? Are you here all by yourself? What a brave boy you are."

Not fooled for a second, Jonathan complained louder.

"How about if I just listen to your heart, how would that be?" I reached for his chest with my stethoscope, and Jonathan lifted his twisted, stiff arms to bar my hand.

"UUUH!"

I tried again to fit my stethoscope in between his arms, and he pushed me away again. Finally I just reached around from underneath his arms and placed my stethoscope against his chest, reassured to hear the swift, regular beat of his heart below. I shifted my stethoscope to ensure that his lungs were still clear. "Well, Jonathan, that's it. You were such a brave boy, and now I won't bother you anymore."

Finally I knew. I could understand why she wasn't ready to let him go and what she was afraid to lose.

I was shocked by the sheer size of the X ray. In contrast with the miniature ribs and hearts on the miniature films on either side, this adult-size film appeared gargantuan. The large breast shadows were startling.

"Who is *this*?" someone asked.

"That's Leslie. She's our fifteen-year-old with congenital HIV. She's a rule out pneumonia," Danielle reminded us. She was admitted to the hospital because her primary care physician was concerned that her sinusitis might be the beginning of a pneumonia.

"Well, that looks clean to me," the radiologist said. "I don't see a pneumonia there."

Leslie was remarkably healthy considering her age. She was

virtually disease free and had never had an AIDS-defining illness. She had a CD4 count into the four hundreds, well above the two hundred limit defining AIDS.

But at fifteen, Leslie also had a ten-month-old daughter and a non-HIV–infected sixteen-year-old partner. Her partner was aware of her status, we were told, and they lived together with the baby in his family's home. This was a stable relationship, or at least as stable a relationship as a fifteen-year-old and sixteen-year-old could maintain.

"That's disgusting! How could she have done that? How could she, knowing that she had AIDS, go and have a child? That is just nauseating," Danielle had said when she found out. "Is the child positive? Was this planned?"

"As far as we can tell, the baby is not, and I think this was a planned pregnancy. But we have to ask her because I'm not certain," Karen said. "But this is an interesting question. Why *do* you think she did it?"

"I don't know. It's so disgusting I just can't imagine how she could have possibly done this."

"Well, maybe she was just looking for some love and permanence in her life," Karen said. "There are all those studies saying that many of the inner-city girls who get pregnant are actually looking for someone to love them."

Danielle thought for a minute. "Maybe she knew that she would die young and thought that this might be her only chance to have a child. It used to be that people got married at thirteen and had children by sixteen because they died in their twenties."

"There is a high correlation between getting pregnant at thirteen and fourteen and childhood sexual abuse. Maybe there were social issues that prevented her from making a good decision," I said. "Or maybe since she has been so healthy, she's in denial and doesn't really believe she is sick or could pass it on to her child."

"It's interesting, because do you guys know the risk of transmission of HIV if you don't breast-feed? It's only twenty or twenty-five percent, and if you take AZT during your pregnancy, that drops to only eight to ten percent. That's pretty low," Karen said. "If you and your partner carry the CF [cystic fibrosis] gene, your chance of passing on CF is much higher than that. We've all seen the families where multiple siblings have the disease. And CF is just as certain death as AIDS. In fact, with AIDS, kids can live ten or fifteen years without any illness, and now with the protease inhibitors, there is real hope for those kids. CF kids are sick almost from the git-go. But since CF is genetic, we don't even think twice about letting them have kids."

I never found out why Leslie chose to have that baby or even if it was planned. With only sinusitis, she was the healthiest patient on our service, and we were never called in to see her. She stayed for observation for another two days until everyone felt comfortable sending her home. I met her only as she was getting ready to leave. Her boyfriend was in the room with her. In the process of dressing, Leslie ran out of her room wearing only a baggy plaid flannel shirt. She caught my classmate, who had admitted her, as he walked by her door. "Can you get me that note for school?" she asked.

"Sure, I'll get it right away," he said.

"I gotta get back to my baby, so hurry up, okay?"

My six weeks of pediatrics passed quickly, and I was eager to leave for my luxurious two-week Christmas vacation. From talking with my classmates during Patient-Doctor, I knew that we all were more than ready for a break from the long hours, overnight call, and performance anxiety of the wards.

Carlos and I thoroughly enjoyed our vacation. We each went home to our own families for the first week, and I spent my days working hard on this book. We met in Boston for New Year's, which we celebrated with a quiet evening at home and a bottle of champagne. The following morning, we drove to Vermont for four days of cross-country skiing.

I have been an avid cross-country skier since my childhood in western New York. Trails were within a fifteen-minute drive from my home, and we generally had ample snow. But Carlos, growing up in California, had never been. I had always dreamed of a partner who would be able to ski with me, and I was touched that Carlos was willing to make the effort to learn. He borrowed my brother's skis, and I was impressed by how quickly he was able to catch on to the technique. We skied on snow-covered frozen ponds lined with woods and tumbled cottages during the days, and we had long dinners at small inns every evening. It was a relief to be in nature, breathing fresh winter air after the stale and stressful atmosphere of the hospitals. After spending exhausted evenings together in Boston, where we were barely able to sit through a video, we found it refreshing to be rested and to renew the energy in our relationship. Although it was only a few days, I felt my spirit was renewed. I returned, if not quite eager for the responsibilities of the hospital, excited to begin my rotation in adult medicine.

Internal Medicine

I had waited for my internal medicine rotation all year long. "It's the high point of your medical school career," a classmate ahead of me had said. Internal medicine, with the help of specialist consultants, covered a wide range of health issues from cancer to kidney disorders, from pulmonary disease to autoimmune illnesses. My classmates and I had spent second year learning almost exclusively about general medicine. We had a better sense of the diseases, the appropriate tests, and the possible treatments. Of all the general areas of medicine, internal medicine was reputed to have the most patient contact. I was sure this was what I wanted to do with my life.

Yet by mid-January, only two weeks into a three-month rotation, I was miserable. I had trouble connecting with my patients and with my team. I wasn't excited by the cases I saw or the people I met. Was I burned out? Was it possible that I didn't like internal medicine? The contrast with my intense involvement in pediatrics was profound. Surely I hadn't been this miserable during pediatrics, had I? With residency applications due in only five months, the deadline for choosing a specialty rapidly approached. This doubt was troubling.

After my experience in pediatrics, I was amazed by the striking absence of HIV- and AIDS-related complications among the adult inpatient population. During pediatrics the team always had at least one HIV patient on the service, and often more. But during my three months of adult medicine I saw only one HIV patient.

One of my senior physicians thought this was related to the advent of the new AIDS drugs. "You know, it's amazing. With the new protease inhibitors and better prophylaxis, AIDS has become largely an outpatient problem. Home care has gotten a lot more advanced too, so people who used to come into the hospital for procedures are now getting them in the office or at home. It's amazing," he said. At least for the adults, AIDS didn't seem to be quite the death sentence it once was, unless they were dying at home.

March, the third and final month of my medicine rotation, was probably the hardest month of the year. I spent the month working on an inpatient ward at a Veterans Affairs (VA) hospital. I felt suffocated by a thick depression. I fought back tears daily, but the reasons for these tears were trivial: There were no new admissions; I was sick of reading; I gave a disorganized presentation; I answered a question incorrectly. Normally I weathered these daily annoyances with a much better attitude. Almost every call night I phoned Carlos late at night from the hospital on the verge of tears.

In the middle of the month, as I learned my way around the hospital and became more comfortable with my team and my patients, the depression lifted. I began to enjoy being in the hospital again. I knew I had a week of vacation coming up at the end of the month, and the knowledge that I would soon escape the hospital propelled me through the final weeks.

Toward the end of my month at the VA, to my surprise, I

began to feel confident about doing procedures for the first time. Although I had spent the previous eight months learning and performing various minor procedures, I was always petrified. I feared the pain I inflicted and the extra stick if I missed. There was debate over whether or not to give lidocaine as a local anesthetic before placing an IV or drawing an ABG. The sting of the anesthetic as it took effect was more unpleasant than the needle stick itself, some argued. I used the anesthetic. It might not help the patient with his pain, but it certainly helped me with mine.

But all of a sudden I wasn't afraid of the pain anymore. These men had certainly suffered much worse than my small needle stick. Stoic and macho, they surrendered without flinching to the various procedures. They helped me separate the pain from the procedure. With this new confidence, I could be a little more aggressive, a little more assertive. Then I began hitting more of my blood gases, more of my IV placements on the first try. My hands shook less, although I have to admit that my stomach was still uneasy. It was a tremendous relief finally to feel competent after my yearlong struggle with inadequacy. I wouldn't exactly call myself good, but I finally gained the capacity to practice and improve with confidence.

The VA was another world, unlike anything I had ever experienced before. Although the hospital was only twenty-five minutes outside Boston, the two buses I took to get there could stretch the commute to more than an hour. The hospital was a mammoth cluster of buildings set just at the edge of the nearest community. As I approached the back side of the hospital on the bus, I first saw the Georgian brick tower rising up from the random conglomeration of brick and concrete buildings. The white wooden cupola housed a large metal bell, and the black pointed roof and weather vane nearly touched the low winter sky.

The main emergency entrance opened into a large, dimly lit sitting area furnished with shabby mauve institutional sofas and love seats arranged to form repeating squares of conversation space. Between seven in the morning and six at night I invariably stumbled on a group of older vets sitting on the couches, wooden canes resting at their sides, or sitting in their wheelchairs, talking. They were mostly aging white men with thinned gray hair and faded blue tattoos. A few TVs were always tuned in to CNN or game shows, but nobody watched, and the volume was always turned down very low.

One inpatient medicine ward was on the fourth floor of the hospital, above the waiting area. The small, cramped rooms generally held three beds. Shabby green curtains offered makeshift privacy, but these were rarely drawn by the patients. They preferred to hang out together, and often we interrupted card games and conversations when we came in to do exams or for rounds. The other inpatient general medicine ward was in the newer wing of the hospital. In this ward the rooms were significantly larger, and each man had several square feet of brown linoleum floor to himself.

But the heart of the hospital was undoubtedly the smoking shacks. The no-smoking institution provided makeshift rooms designated for smoking, since virtually every patient was an avid smoker. These rooms were unheated, unfurnished "sunporches" glassed in with industrial brown metal window frames and concrete floors. If we couldn't find the patients in their rooms, we usually found them here.

Because the government was in the process of reorganizing its hospital system to improve efficiency and cost-effectiveness, the administration had closed many of the smaller hospitals. Rather than fire the nurses who had worked at those sites, the government relocated them. One call night I had seen a nurse all day long. She was already there at 7:30 A.M., when I arrived, and at 10:00 P.M., when I came down to check on a patient, she was still there. "This is a long day for you," I commented.

"Yeah. I worked a double today. Two eight-hour shifts. They closed the other hospital I used to work at, so I had to come here. But I live a hundred twenty miles away, on the Cape. With traffic and everything, I prefer to work three double shifts a week on the weekends to cut down on the commute," she said.

"That's incredible," I said. "Does everyone do this?"

"Oh, no. You need a special evaluation. If you've called in sick more than five times in a year, you can't get this. And if you call in sick on a double, then you're out because it's too hard to cover. It's a tough schedule. By Sunday I'm exhausted and crabby with the patients, even though I don't always mean to be. Thank God, I've never made a med mistake yet," she said, and crossed herself. "But it's my big fear that I will one of these days."

Bizarre things always seemed to happen on the wards of the VA. I'm not sure if the patients were more unusual than elsewhere, the system more convoluted, or some combination of both, but humorous stories abounded at the VA, stories that could never be generated elsewhere.

Carlos had also done his medicine rotation at this VA, and one call night his intern thought she smelled smoke. "But she was always worrying about something or another, so we didn't take her seriously. Then, a few minutes later, we saw all the nurses rushing into one of the rooms down the hall. Apparently one of the patients in three-point restraints had tried to set himself on fire. The patient in the next bed had turned him in. By the time we got there, things were under control. A little while later I walked by the nurses' station, and I overheard one of the nurses saying to another nurse, indignantly, 'That's it. That's the last time I give the patients matches!' "

On another classmate's call night a patient on her team demanded transfer back to the psychiatric hospital he lived in. He was yelling in his room and disturbing other patients on the floor. From a medical standpoint the patient was ready to return to the VA psychiatric facility, but as usual things were arranged slowly, and he had already lingered on the medicine service for an extra four days. Although he was not her patient, my classmate finally went in to placate him. After explaining there was nothing she could do about the situation, she turned to leave. As she walked out of the room, she felt something in her back. She turned around and saw the patient stabbing her with the knife from his dinner tray. The orderlies rushed in to restrain the patient, and my classmate was not hurt. The patient was back at the psychiatric facility first thing the following morning.

On my service we had a seventy-year-old alcoholic with poorly controlled diabetes. His left leg had been amputated just above the knee because of a complication of the diabetes. He then contracted necrotizing fasciitis, the same flesh-eating strep infection that my surgical patient Roger had, although this man's infection was less severe. This patient had already been in the hospital for weeks awaiting surgery to cut off the diseased and dying tissue and bone. But for some reason never entirely clear to us, he was delirious. One morning the residents and I walked in for morning rounds, and he seemed more alert than usual. He was watching TV, and he wished us good morning when we came in, something he had never done before.

Encouraged, the intern decided to recheck his mental status. "What's your name?" he asked.

"Jay Leno."

"No, really." My intern smiled. "What's your name?"

"Jay Leno!" he said, his blue eyes shining indignantly.

"So, you have your own night show?"

"What?" he asked, confused.

Many of my patients at the VA faced end-of-life issues. Through my hospice experiences, I was already familiar with some of these issues. But my hospice patients came to me because they knew they were dying. At the VA I diagnosed my patients with terminal illness. For the first time one of my patients died unexpectedly. I sat with families as they struggled to decide what to do for their ailing relatives. I helped make recommendations about whether or not to choose aggressive resuscitation. Some of our patients were not able to make their own choices and had no families to help them. I felt responsible for helping my patients make appropriate choices, but I didn't always know what they were. More than ever I felt the skew in the patient-doctor power balance.

Tiger

Mr. Harvey was in his early seventies. Every morning we arrived to find him lying in his bed, awake, with the dingy hospital green curtain partially pulled to separate him from his roommate. He peered beyond the edge of the curtain, brown eyes fixed on us as we entered the room. Mr. Harvey was a long, thin man. Dark silver hair framed his face, and he had a small mouth with thin lips easily lost in his long face. The white cotton woven blanket was pulled up to his midchest, and his arms lay on top of the covers. His long fingers toyed with the folds of the cotton blanket. His hands were wasted from illness. The muscles and flesh had receded deep into the crevices between the bones. His fragile skin, dotted with the brown freckles of age, stretched across his hands, emphasizing the long bones extending to his wrists in bas-relief. A gold wedding band hung loosely on the fourth finger of his left hand, slipping up and down as he gestured to help make his point.

Twenty years before, Mr. Harvey had suffered a stroke affecting the Broca's area of the brain, which organizes the mechanical formation of speech, while leaving the area controlling comprehension and creation of language intact. He could understand everything but was almost unable to communicate.

This was Mr. Harvey's eighth hospitalization for abdominal pain in the last eighteen months. Each time he presented with symptoms caused by partially blocked bowels. His abdomen became distended and painful, and he could not pass stool. But no one could pin down the cause for all these episodes of obstruction. After IV hydration and stopping food by mouth for several days, the obstruction spontaneously resolved, and Mr. Harvey returned home again until the next episode.

"Good morning, Mr. Harvey. How are you feeling this morning?" the senior asked. Adam, our senior, was calm and easygoing. He had a sense of humor about our role, and I appreciated how seriously he took me and my opinions. He often sat down and chatted with me, sometimes about a particular patient but often just about hospital and team life in general.

"Wah, I behind mel jwah. I, I, I," Mr. Harvey said. He fixed his brown eyes on Adam, willing him to understand. "Behind I me noll." He realized we didn't understand. Frustrated, he pointed with his finger toward his belly. "Jesus Christ. Jesus Christ!"

Adam pulled down the blankets and raised the brown hospital pajama top to reveal Mr. Harvey's swollen belly. The smooth, pale skin stretched over his distended bowels. A shiny pink scar extended vertically from his navel to his groin.

"Jesus Christ!" He gestured with his hands to his belly. "Oh, Jesus Christ."

Adam pulled his shirt up further to listen to his heart and lungs. The thin skin rose and fell over each rib. As Adam leaned over Mr. Harvey and laid his stethoscope over the valley between the second and third ribs, his navy blue paisley tie gently fell forward and slipped across the folds of the worn white blanket.

To define the mechanism of his obstruction better, Mr. Harvey was scheduled for a CT scan of his abdomen. The study would allow us to visualize the soft tissues and bony structures of the abdomen in detail. We told him he would go down to radiology later that morning.

"Jesus Christ. Jesus Christ." Mr. Harvey looked at us and shrugged with his hands.

That afternoon Dr. Mansfield, the senior physician; Luis, one of the interns; and I went to radiology to review the CT images of Mr. Harvey's abdomen. I was a little afraid of Dr. Mansfield the first time I met him. He was one of the more prominent physicians at the VA and shared responsibility for the residents and the medical student rotation. He was always quick to voice his opinion during large-group meetings, and he certainly didn't mince words. He was average height and wore his red-brown hair short with a goatee. He had a large collection of multicolored, boldly patterned 1970s ties that he had probably acquired sometime during his residency. Even with the advent of retro, they couldn't quite pass for high fashion.

But I soon adjusted to his style. Sometimes I thought that he was unduly harsh to me, my teammates, or other clinical staff, but I was also able to appreciate the moments when Dr. Mansfield was caring and supportive. Ultimately, of all the senior physicians I worked with, he took the most care and interest in my education. I appreciated the attention and extra teaching. We even occasionally got into friendly arguments over how to manage a particular patient, and I enjoyed doing the extra research to prove my point or, more often, to prove his.

Dr. Mansfield, although deeply involved in the management of each of our patients on the team, took particular care with Mr. Harvey. He was perplexed and concerned about the recurrent obstructions without an identified cause, and he was anxious to get the results of this CT scan. The radiologist to whom we took the films for evaluation was another of the better-known figures in the hospital. If Dr. John had read the film, there was no question about the interpretation.

Dr. John was a round sort of person. He was balding, and his thick brown mustache formed a broad rectangle between his nose and lips. He stood back from the light board with his arms folded across his chest as he studied the films. "Look at this here." He pointed to a large protuberance on the left side of the prostate on the CT scan.

"Look at this; look at his marrow space here." Dr. John pointed to a fluffy white-gray density in the center of the bright whiteness of the dense cortical bone outlining the pelvis. "This should be black." He perused the images again. "This is worrisome. Metastatic prostate cancer can do this."

He rifled through the thick jacket of X rays and pulled out a few more images from this study. "Look at this vertebra here; look at this patchiness. This is very troubling."

The radiologist pulled out a CT study of the abdomen from a year earlier to compare. Although the marrow spaces were fuzzy gray then too, the fuzzy patches were smaller. There had been much more blackness to the center of the bones.

"This makes me very worried. Of course there are other things that can do this, but the more I look at these films, the more convinced I am that this man has metastatic prostate cancer. I can't say for certain, of course, but that is my leading diagnosis."

As we walked out of the radiology suite, Luis remarked, "Shit. I really hope it's not metastatic cancer." He thought for a minute, recalling Mr. Harvey's emaciated frame. "He certainly has the look about him."

The next day Mr. Harvey's prostate specific antigen (PSA), a marker for prostatic cancer, came back at 1173, when the upper limit of normal is 6.

Later that afternoon Luis met with the oncologist, Mr. Harvey, his wife, and his daughter to discuss his treatment. Mr. Harvey and his family decided they would not opt for any heroic measures, and they wanted to minimize invasive procedures. He wanted to go home as soon as possible. He and his family also decided to change his code status from aggressive resuscitation to do not resuscitate, do not intubate (DNR/DNI).

Before we realized Mr. Harvey had metastatic prostate cancer, we had initiated a consultation with the gastroenterology (GI) service, which now also followed his case. The GI team of doctors wanted Mr. Harvey to have a colonoscopy to ensure there were no malignant masses in his bowel to cause the intermittent obstruction. The doctors remained concerned about a second, independent malignant process inside the bowel despite the evidence of prostatic pathology to explain all his symptoms.

I wondered what difference it would make to his care plan if we found a separate malignancy. Would Mr. Harvey choose the complicated surgical resection? I thought he would not. But we pushed these issues aside, since the gastroenterologists wanted the safe and relatively painless procedure and Mr. Harvey's family had consented.

On the morning of the colonoscopy, we arrived in Mr. Harvey's room to find him particularly uncomfortable and unhappy. "Jesus Christ." He pointed to his belly and then to the door. "Jesus Christ." He pointed to the door again.

Adam lifted Mr. Harvey's shirt, and his pale skin was stretched even more tightly across his distended abdomen. He was markedly worse than the morning before. To rid his bowel of stool in preparation for the colonoscopy, Mr. Harvey had drunk a gallon of GoLYTELY, a pineapple-flavored preparation that cleanses the colon. While he managed to down nearly two-thirds of the jug, he had not yet had a bowel movement to pass the fluid and stool. The gallon of GoLYTELY still sat trapped somewhere inside his bowels.

"Jesus Christ. Oh, Jesus Christ!" he moaned.

Later that afternoon we received the results of the study. The colonoscopy was normal, except for the stool and GoLYTELY trapped in his large intestine.

During my month at the VA we had several patients at different stages of the dying process. Much as we wanted to treat them thoroughly, appropriately, and sensitively, the issues we encountered were complex. For patients who asked for "no heroics," how should we decide what was too much? How hard should we look to find a diagnosis? In each case it was hard to know whether we did the right thing. Most of the time I think we did, but sometimes I knew we didn't.

Mr. Marino arrived late on a call night. "Are you sure you want this one?" Adam asked me. "There's no reason you need to take this. It sounds like it will be pretty depressing. The ED people say he looks bad."

Mr. Marino had been diagnosed with lung cancer two years before. He had refused a workup to identify the type of lung cancer and treatment at the time, and he was coming to the hospital now with increasing shortness of breath and hoarseness. A CT scan the week before showed massive spread of his original tumor in the right lung to both lungs. The tumor pressed on his trachea and partially obstructed his right main stem bronchus, one of the primary passageways for air. He had what was most likely a metastatic lesion in a vertebra of the midback.

"Yes. I definitely want this one," I told Adam. After working with hospice, I had grown to love dealing with patients confronting the end of their lives. In the hospital few caregivers seemed to be aggressive about involving their dying patients with hospice, and I thought I could make a difference, both for the patient and for the team, by lobbying strongly for my own patients. But most of all, patients at the end of life often appreciated extra time to talk, and as a medical student I was the only one on the team with extra time. I definitely wanted Mr. Marino.

So at nine-thirty Luis and I looked for Mr. Marino to take a history and do a physical exam. Luis was one of the best interns I came across during my year. He had deep brown eyes and a gentle manner that his patients loved. He had an incredible fund of knowledge. I could ask him anything, and he made time to explain the answers. Luis often went out of his way to find me and warn me about some detail I was likely to be quizzed on during rounds.

The two of us went off in search of Mr. Marino. We finally found him in Mr. Harvey's room. He was Mr. Harvey's new roommate. Mr. Harvey, cloistered in the dimness protected by his curtain, watched us as we walked into the room. "Hey, Mr. Harvey," I called to him as we walked into the room, "how are you doing?"

He lifted his hands and shrugged. "Jesus Christ."

The brightness of Mr. Marino's half of the room contrasted starkly with Mr. Harvey's dusk. Mr. Marino sat in a wheelchair, his back to us. The clear, thin ribbon of the oxygen tubing wound around his dark green hospital wheelchair toward the gray hair in the buzz cut. The ribbon snaked out of view to the front of Mr. Marino and then reemerged as two distinct translucent bands just above his shoulders, one on each side. The two bands synchronously wound around his ears and disappeared again as they escaped around the angle of his cheekbones.

Another man stood near the door to the room. Intensely involved in a conversation with the nurse, he made sure she carefully recorded Mr. Marino's medications and dosages. The man was fairly young; his hair was still dark brown, and the ribbing of his kelly green polo T-shirt strained across his biceps. Yet another man sat in an orange vinyl chair next to the wall. He had dark gray hair and large glasses that were tinted maroon. He wore a blue baseball cap, a nylon jogging suit, and bright white sneakers. He sat quietly in his chair, watching as the younger man talked with the nurse.

I moved around to the other side of the wheelchair to introduce myself to Mr. Marino. He had fuzzy gray eyebrows dwarfing his light blue eyes and a rounded, long nose with a gray-brown mole just to the left of the bridge. He wasn't wearing a shirt, and despite a weight loss of sixty pounds over the last eight months, his large belly still sagged onto his lap. He had a faded blue tattoo on his right forearm, and he wore a pair of old navy blue sweatpants. "Mr. Marino? I'm Ellen, and I'll be part of the team taking care of you. This is Luis, who is also part of the team."

"Hi," he whispered. He waved at us dismissively. He raised his fluffy eyebrows and looked away from me, toward the man sitting in the vinyl chair. As we talked, it became clear that the two men were his sons.

"How are you doing tonight?" I asked him.

"I'm pissed off!"

"Why? What's bothering you?"

"They lied to me!" he whispered angrily.

"What do you mean?" I asked.

He looked at me for a minute without speaking, as if he expected me to know. "They said my lungs are worse! But before, they said there was nothing there. If I hadn't stopped drinking, I wouldn't be here!" he whispered. I knew from his records that Mr. Marino was an alcoholic who continued to drink.

"When did you stop drinking?" Luis asked.

"It's probably been over a month since he last had something to drink," the older son said. "Right, Tiger?"

"Yes!" Mr. Marino said.

"And you think that if you had kept on drinking, you wouldn't be sick in the hospital now?" I asked.

"Yes!" He looked at me, fuzzy eyebrows raised over his blue eyes, incredulous that I questioned his theory.

"So, you're upset because you thought your lungs were getting better, but now you find out that they are actually getting worse?" Luis asked.

Mr. Marino turned in his wheelchair to look at Luis directly. He raised his fluffy eyebrows and pointed at him. "Wouldn't you be?" he whispered angrily.

"This news is very upsetting. You have every right to be angry," I told him. "But in order to help you tonight, we need to understand what has been happening with you. Why don't you tell me what has been going on for you over the past few days and what brings you to the hospital tonight?"

"I can't talk!" he whispered angrily.

"I can see. How long has this been going on?"

"I don't know. Couple of days."

"Tell me how things have been for you over the last week. Have you had a cough or fever?"

"I've been coughing . . ." he whispered, and then got frustrated. He turned to the younger son still talking to the nurse and brusquely gestured for him to come over.

"What's that, Tiger? We always call him Tiger. It's his nickname," his son explained.

Mr. Marino gestured to me and Luis. "Tell them!" he whispered.

"We were just wondering how he has been over the past week and what exactly has brought him into the hospital," I told the younger son.

"Well, I don't exactly know. I don't see him every day, but the visiting nurses do. He lives in an apartment in a senior housing system. He's been getting sicker and sicker over the last month. It used to be that he would go out to the grocery store and get out of the house. But he can't do that anymore."

"But how about the breathing? How has the breathing been?"

"Well, he's been using the oxygen. But he's getting short of breath."

"I used to walk with a walker. But with this goddamn oxygen, I can't carry the oxygen and the walker. I can't even get to the goddamn bathroom anymore!" Mr. Marino interjected.

"But has he had a cough or fever?"

The older son chimed in. "He has been having trouble breathing now for a long time. And he's had a cough for a month. But he was really doing okay until just this month."

The conversation was going nowhere, and I felt totally out of control of the situation. Neither the sons nor the father seemed able to give an appropriate, relevant history. At times there were three conversations going on at once. While I tried to decipher Mr. Marino's whispers, Luis talked with the older son, and the younger son was immersed in his conversation with the nurse. I was distracted by the extraneous conversations and lost the thread of my own.

"I'm just trying to understand exactly how things have changed and what made you decide to come to the hospital tonight," I told Mr. Marino.

"I can't talk! I can't live like this! What do you expect me to do? I live by myself. How can I stay at home if I can't use the phone and call someone to help me?"

"Yes, we'll do our best to straighten your voice out. But I need to understand what has been happening with you over the past several weeks so we can give you the best care possible. Have you been able to eat and drink over the past few days?"

"No, he can't eat anymore," the younger son told us. "He's been having trouble swallowing." I worried that this inability to swallow represented spread of his lung cancer.

"Okay, Mr. Marino. Are you having pain anywhere else in your body? Are you having back pain or headaches?" Luis asked.

"He's been having some back pain recently," the younger son said.

"It's from all this goddamned sitting. Your back would hurt too! I just got to lie down," Mr. Marino whispered. He motioned to his younger son.

"What's up, Tiger?"

Mr. Marino pointed to the bed.

"You want help to lie down?"

Mr. Marino gestured angrily again toward the bed.

"Okay, Tiger. Let's get you up. Do you want to tell them why we call you Tiger?"

Mr. Marino sucked in a big breath and gripped the handles of the green wheelchair. He pushed with his hands as he struggled to get out of the chair. He lifted himself a few inches, fell back, and then pushed himself up to nearly standing. His son caught him and supported him as he took the two steps to the bed. Mr. Marino dropped himself heavily onto the bed, and he sat gulping for air.

"When we were just kids, a dog came and bit my dad. But instead of giving up without a fight, my dad bit back, and he wouldn't let go. So since then we've always called him Tiger. We've never called him Dad."

"Here, Tig, why don't you have a sip of your orange juice?" his older son said, and passed him a cup with a plastic straw from his nightstand. Mr. Marino took a sip and began coughing and choking. He brought up orange-stained phlegm. His pale face turned bright red. He coughed for a few minutes until finally he regained his breath. "Goddammit." This was a bad sign. His swallow was not coordinated, and the orange juice had made its way into his lungs.

As we began the physical exam, the older son said, "You know, he's been having accidents. He's wearing diapers now."

"Yes," Mr. Marino whispered.

"Bowel and bladder incontinence?" Luis asked the son.

"Yes, both."

Mr. Marino's physical exam was fairly unremarkable despite his very advanced disease as evidenced by the CT scan and his poor physical condition. He had some prominent lymph nodes in his neck and possibly a mild decrease in strength of his quadriceps muscles. His lungs sounded clear without evidence of pneumonia or the wheeze of emphysema. Mr. Marino was worn out by his evening in the emergency department and his interview with us. By the time we finished the neurologic examination, he had already nodded off.

We explained to the sons that in all likelihood the progressing lung cancer caused his symptoms. We would call an oncology (cancer) consult in the morning to decide what therapies we could offer to make him more comfortable and better able to manage at home. His sons had watched him decline over the last month and were prepared for this news.

"We just don't want any heroics," the younger son told us. "He refused evaluation of his tumor before because he was afraid of the chemotherapy. He has seen many of his friends die, and they all felt terrible on the chemotherapy and died shortly thereafter. He doesn't want anything to do with that."

After we left the two sons, Luis and I returned to the main work area. Now, close to 11:00 P.M., the workstation was quiet. We paged Adam, and he joined us within a few minutes to review Mr. Marino's situation.

"So, what do you think, Ellen?" Luis asked me after I had summarized Mr. Marino's history and physical for our senior.

"Well, I don't know. He doesn't seem to me to have a pneumonia or acute infectious process, and I was also unimpressed with his exam for asthma and emphysema. But I would treat him for an emphysema flare anyway with antibiotics and nebulizers and low-dose steroids to ensure we maximize his respiratory status. Otherwise I think we need to go down the route of palliative care and possibly even hospice."

"I agree with you about his pulmonary situation, but I am worried about spinal cord compression by metastatic spinal tumor," Luis said. "That is one of three oncologic emergencies, and anytime anyone comes in with new onset bowel and bladder incontinence, cord compression has to be the first thing that crosses your mind. These people need an urgent MRI and steroids and then radiation therapy or they could be paralyzed within the week. I am also very concerned about his trouble swallowing."

"How concerned are you really about cord compression? Do you think this guy really needs an MRI tonight?" Adam asked Luis. Obtaining an MRI that night was not a trivial matter. It was the only diagnostic study capable of evaluating for cord compression, but the only MRI machine in the local VA system was at another hospital. Mr. Marino would have to be taken by ambulance to another hospital thirty minutes away to get an MRI.

"Well, I don't know. Recent onset bladder and bowel incontinence makes me worry. But otherwise he had a nonfocal neurologic exam, which makes me less concerned. There is some proximal muscle weakness, but it is only very slight, maybe four plus out of five. So I guess that overall I am less concerned about cord compression. He's tired and very upset tonight. I think we could probably wait until tomorrow," Luis said.

"And what do you think, Ellen?" Adam turned to me.

"Well, I don't really know. I've never heard about cord compression before, so I don't know how fast it progresses or really anything about it. But if you feel comfortable letting him wait, I would too."

"I agree. It sounds like the level of concern isn't that great," Adam said. "I think he can wait and have an oncology consult first thing in the morning."

"And how about a dose of steroids tonight?" Luis asked him. Steroids during the initial phase had been proved to limit pressure on the cord and preserve function.

"Well, if our suspicion of cord compression is low enough that he's not going to have an MRI tonight, then I think we can wait on the steroids as well."

We spent another half hour discussing Mr. Marino's plan and medications and signed all the orders before Adam was paged again.

Luis sat finishing his note and some other orders while I began my admission note. "I'm really worried about this guy, Ellen. This makes me very uncomfortable. I know an intern should never sleep on a cord compression. If we were at the Brigham [another of the Harvard hospitals], then this guy would get an MRI tonight, no question. But here it's not so simple. We'll just have to be sure to call the oncology consult first thing in the morning, so he gets seen early."

Luis, now caught up with his work, left to catch a few hours' sleep before he would be paged again. I sat for several more hours writing my note and reading a little about lung cancer. I knew Dr. Mansfield would expect me to have evaluated his sputum for any evidence of bacteria to justify our antibiotic therapy, but each time I went into Mr. Marino's room, he was fast asleep, and I didn't have the heart to wake him to cough up some phlegm into a sterile cup. After reading about cord compression and its symptoms, I too was worried about Mr. Marino.

The next morning I woke early to get a sputum sample from Mr. Marino so I could have the results by the time of morning rounds to present to Dr. Mansfield. I found Mr. Marino awake in his bed, still shirtless with his navy blue sweatpants. He was in a foul mood.

"Good morning, Mr. Marino. I'm Ellen. Do you remember me from last night?"

"It's so goddamned hot in here! I couldn't sleep all night. Jesus Christ!" he whispered, echoing his roommate, Mr. Harvey.

"How was the breathing overnight?" I asked him.

"How do you think it was? It was terrible!"

"I know you've been very uncomfortable. But was it better, worse, or the same as when you came in last night?"

"Maybe a little better. The oxygen helps."

"Okay. And how is your back?"

"It's killing me!"

"Did you take any Tylenol?"

"Yes, but it didn't help."

"We wrote you for some additional pain medicine if the Tylenol didn't work. Did you take anything else?"

"No."

"Well, there is no reason you should be in pain from your back. We can give you medication for that. You just have to be sure to tell us when the medicines aren't working for you."

I listened to his heart and lungs. "Do you think you could cough up some phlegm into this little cup for me?" I helped him sit up, and he coughed up a tablespoonful of tan phlegm and spit it into the base of a small plastic container. I took the container, told him the oncologist would be by to see him later, and headed off for the laboratory.

In the small laboratory the white porcelain sink was stained pink and purple from years of fixing solutions, and an old microscope sat on the black bench. Old slides and used pipettes littered the countertop, and a jar of some unidentifiable bodily secretion sat in a plastic container with a red cap on top of the tall garbage can marked BIOHAZARD. I quickly found the much-thumbed photocopy of instructions on gram staining to evaluate for bacteria in sputum or other secretions. The last time I'd made a gram stain had been well over a year earlier, and I didn't feel very confident of my abilities. I made my slide and tried to evaluate it, but I left the container with the tan phlegm and the slide on a piece of paper towel labeled "Marino, 3/97" in case I needed to come back and repeat the study.

After I finished the slide, I tried to page the oncology consult service, but the oncologist was not yet in the hospital. I would have to wait until after rounds. When I paged back later that morning, I reached the oncologist easily and explained our concern about cord compression. Although the oncologist thought there might also be a role for chemotherapy or radiation therapy to slow the progression of disease, he also thought the objective of his meeting with Mr. Marino and his family would be to discuss how aggressive they would like to be with his health care over the next months.

I went to let Mr. Marino know the oncologist would be by to see him later that morning. As I walked in, Mr. Harvey was on a stretcher being taken away for another procedure. As I squeezed by the gurney to get into the room, Mr. Harvey looked at me and lifted his hands. "Jesus Christ!" he said, and rolled his eyes.

I found Mr. Marino sitting in the green vinyl armchair next to his bed.

"What is this goddamned place?" he whispered as soon as he saw me. "They got me up into this damned chair and just left me here, and I can't get out of the damn thing. What do they think they're doing? The VA is not what it used to be. Jesus Christ!"

"Would you like some help getting back to bed, Mr. Marino? Tell me how I can help you."

He pointed to the ground in front of him. "Stand here!" he whispered. I moved to where he motioned me and watched as he braced his hands on the arms of the chair. He pushed with legs and arms, and his face turned red with the effort. He hoisted himself a few inches out of the chair and then fell back again. He made another effort and managed to raise himself a few more inches out of the chair before falling back. I tried to catch him as he rose out of the chair, but my small frame was not sufficient to support his bulky body. A nurse helped me push the chair a few inches closer to the bed so he could shift from the chair to the bed without walking. Mr. Marino pushed with his arms to lift himself out of the chair. He wavered for a few precarious moments with his legs still bent at a ninety-degree angle, but finally he rose to standing. He immediately turned around and sat heavily on the bed. Watching him get out of the chair, I was even more concerned about cord compression than I had been the night before. His weakness, an early sign of compression, seemed more marked than I had recognized the night before.

"And the goddamned TV. Where the hell's the remote? I can't reach it!"

I found the white remote dangling from its cord, which had become tangled in the protective rails of the bed. "Do you want this on?"

"*Yes!*"

I pressed the red on button, and the TV turned on. But when I released the pressure of my thumb, the button was stuck in the on position. The TV flipped through *I Dream of Jeannie* to the news to *Sally Jessy Raphael* and turned off and then turned on again in an endless cycle.

"This goddamned place. Nothing the hell works here!"

"I think this remote control is beyond my skill level, but I'll check with the nurse on my way out. What I really came to tell you was that the oncologist will come by to see you later. He's going to talk to you about all the therapies we have to help make you feel better. We won't be able to cure the cancer at this point, but we can certainly help you feel better. Now I know that you decided that you didn't want biopsy and chemotherapy before, but the oncologist will discuss this possibility with you again today. But the most important thing I wanted to tell you is that you are in charge," I told him.

"Okay." He nodded.

"Well, I need to get going." The TV was still caught in its endless cycle. "I'll find the nurse and see what I can do about that remote control."

I found his nurse in the hallway. "Oh, yeah. He must have the broken one. We keep on switching it from bed to bed as people are discharged," she said, and walked away.

Mr. Marino was our most complicated patient, and we discussed him first at rounds. Dr. Mansfield was not happy with how we had managed him overnight. "You called the wrong consult. You needed a neuro consult last night to evaluate him for cord compression. You should not have waited for the oncologist this morning. Maybe you feel confident enough of your neurologic exam to evaluate for cord compression, but I sure wouldn't have trusted mine. I would have dragged the neurologist out of bed at two in the morning or whatever hour to see this man. And what about the steroids? One dose of steroids wouldn't have hurt him. At the very least he should have gotten steroids last night. This man might not have cord compression. But he might. And you can't dismiss it. It's a serious complication."

I felt frustrated with both Dr. Mansfield's response and myself. Formulating a plan in the middle of the night could be difficult when the patient had complicated issues and when the nurses pressured us to finish the orders so they could start medications. Although most of the time we felt competent to get the patient through at least until the morning, sometimes the right choice was not obvious. In my few short months of medical training, I had learned that what seemed true in the middle of the night was often not true in the light of day. I knew we hadn't felt entirely comfortable with our management of Mr. Marino, but that night I thought we had chosen the most practical route. That morning, however, I wasn't so sure.

I had to leave the hospital early that afternoon to get to the primary care clinic where I worked on Tuesday afternoons. I was unhappy about leaving the hospital. I didn't feel settled about Mr. Marino's situation, and I wanted to stay and meet with the oncologist and neurologist to get a better sense of what we would do for him over the next days and months.

As I waited for a taxi to pick me up and take me to my clinic, I thought about Mr. Marino. He had given up his life to alcohol.

To judge by the history in the chart and his sons' comments, he had been a heavy drinker for his whole life. Now, nearing death, Mr. Marino did not seem to be reaching out to his sons, finding closure in the relationships or finding resolution in his life. He had lived a dysfunctional life, and it seemed to me that he would die a dysfunctional death. I felt sad for him and his family.

Late that evening Matt, one of my classmates at the VA, called me. "Did you get my message?" he asked. "I think one of your patients coded and died this afternoon."

"Mr. Marino? Was it Mr. Marino?" I asked.

"I think so. On the fourth floor?"

"Yeah."

"I really didn't see much. By the time I got there, the whole room was filled with people. It was awful. There was so much blood everywhere. They tried to intubate him, but they couldn't save him. The whole thing was awful. Are you okay with all this?"

I was sad to hear that Mr. Marino had died, but I felt okay letting him go. Watching him interact with his family made me think back to the hospice. Among the patients I had met, there were some who had lived rich emotional lives, very connected to friends and family; they had lived well, and they died well, with or without us. There were those who had lived dysfunctional lives with no insight into themselves or their relationships; for the most part they died poorly despite our efforts. Then there was another group of people who wanted to reach out and with help found resolution and peace in their relationships. Mr. Marino seemed to fall into the second category. My goal as a caregiver was to make him as comfortable as possible and to help him remain in control of both his medical and his emotional life. I didn't believe I had to save him or help him achieve a beautiful death, only to help him go with dignity.

"Yeah, I'm okay," I told Matt over the phone. "I'm sorry he died, but I think it was for the best."

The next morning my team was waiting for me. As soon as

Luis saw me, he came over. "Did you get my message last night? I tried to call you."

"You know, I forgot to check my answering machine, but I knew about Mr. Marino. I talked to one of my classmates who was at the code last night."

"It was just awful, Ellen. I was pretty much the first one there because I was on the fourth floor signing the order to send him to the other VA for an MRI. It was really bloody. There was blood everywhere—all over him, the bed, the floor. He just bled out into his lungs and drowned. We tried to intubate him, but we couldn't get any breath sounds even with the tube in his trachea. He had just decided to be DNR, but we hadn't done any of the paperwork. It was a short code because we saw fairly quickly that it was going to be useless, and he and his family said they didn't want heroics. It was ugly, Ellen. You should be glad you weren't there. Thank goodness we hadn't been quicker about the MRI. He could have died in the ambulance."

Mark, the other intern on our team, found me and told me that last night they all had been concerned about how I would take the news. Mr. Marino's death had been particularly ghastly and troubled them, despite their many experiences with death. The cancer had most likely eroded into one of his arteries, causing him to bleed. As medical students we hadn't seen many people die, and they worried that this terrible death might have been my first. Mark brought in a letter he had written to the family of a patient he had cared for in the intensive care unit as a medical student to let me know how he had dealt with the situation.

Before rounds that morning I went to the student laboratory to clean up the specimens I had left the day before. I walked into the tiny room and found my paper towel among the clutter on the black bench. "Marino, 3/97." I found the cup with the tan phlegm and the purple crusted slide. It felt odd to see the tan phlegm as the last living reminder of a man now dead. It felt somehow an insufficient memorial to step on the pedal of the tall red Biohazard garbage can, lift the lid, and throw the plastic cup into the dark depths within.

Later that afternoon I went with Dr. Mansfield, Luis, and Mark to review Mr. Marino's autopsy. We walked over to the pathology department together. I wasn't sure what to expect. I thought they might have Mr. Marino himself on the dissecting table, and I wasn't sure how I would respond to that. I had briefly thought I wanted to watch the autopsy but then quickly reconsidered. I was afraid it would feel like dissecting in anatomy lab but with all the responsibilities and attachments of the human relationship that, much as I struggled to create it during the first-year course, I was now relieved to avoid.

We entered a cold large room through metal double doors, across the hall from the metal door with a sign in small white capital letters on a black background reading MORGUE. In the room stood three gray metal dissecting tables, each with the same round panel of mobile lights used in the operating rooms. Dull metal cabinets covered the left wall; a metal countertop cluttered with papers, books, soap canisters, and three gray sinks extended the length of the right wall. Two pathologists in green scrubs stood at the third table in the yellow glow of the operating lights. As I walked toward the table, I was relieved to see the yellow light reflected on the flat silver tabletop. The flat grayness of the metal was not broken by the soft hills and valleys of a draped human form.

The shorter pathologist introduced himself as Vikram. He went to the wall of metal cabinets, slid open a door, and brought out a large square metal container, which he placed on the table. He removed the metal lid and revealed a dark, sweetly pungent solution threatening to overflow its metallic boundaries. He reached into the dark, fragrant depths of the metal container, pulled out several pieces of flesh, and laid them out on a metal tray sitting on the table. "We just got these into the formaldehyde a couple of hours ago," Vikram explained. "We can't leave the tissue sitting out for too long. That's why it wasn't all ready for you when you came."

"So, what did you find?" Dr. Mansfield asked.

Vikram looked over toward the taller pathologist, who nodded. "Well, the frozen-section slide of his lungs showed that he had squamous cell carcinoma of the lung," Vikram said. "Otherwise he was completely healthy, by the looks of his organs. If he hadn't smoked, he could have lived a lot longer."

Vikram pointed out the nutmeg brown half-inch sections of liver. "Look at this. You're not going to see a better specimen than this. Even the capsule—it's perfect. Put on gloves, and you can feel the smoothness of the capsule."

I took a pair of gloves and reached out to touch the liver. I felt the smooth, cool firmness of the tissue. The capsule of connective tissue surrounding the liver was glossy.

"And the heart..." Vikram moved his gloved hand to show the red-brown mass of heart. He opened the various vertical cuts to show each of the four chambers. "This is not the typical heart of a vet. He doesn't have any of the wall thickening of chronic hypertension or the dilation and wall thinning of chronic alcoholism. His valves have very little of the calcification of old age."

I felt the tough ropiness of the muscular walls between my thumb and index finger. My gloved fingers slipped across the paper-thin fibrous whiteness of his valve, tracing along the thin white cords attaching it to the red muscle wall.

"But now we get to the lungs." Vikram pointed to the right side of the tray, where one-inch sections of trachea and lymph nodes dissected from his lungs lay. Lung cancer commonly spread to lymph nodes. "Fortunately, lymph nodes are very easy to identify, particularly in a smoker. All the carbon inhaled either from city living and exhaust or, in this case, from cigarettes collects in the lymph nodes."

I saw the faint blue-black speckles dotted throughout the creamy whiteness of the enlarged nodes. I saw the pink of the trachea, and I felt the firmness of the cartilage rings that protect the trachea from collapsing. The speckled nodes felt rubbery and firm, and I saw where the trachea narrowed to half its normal diameter as the speckled cancerous nodes overcame the cartilaginous firmness and compressed its walls. I thought about how these organs had been Mr. Marino less than twenty-four hours ago. He had breathed through this narrowed trachea. This heart had beaten in his chest, pumping blood throughout his body. He had been consumed by these rubbery growths. I had known these organs as a living man. It made me feel creepy to be handling these tissue slices. Yet the organs and the man were strangely disconnected in my head. These inert specimens had nothing to do with the Mr. Marino I knew. They were just specimens.

"And how about cord compression?" Dr. Mansfield asked.

"I noticed you had raised that question. I ran my finger down the entire length of the spinal column, and I wasn't impressed by any narrowing. We really didn't find any lesions that would be responsible for that kind of cord compression either. But, of course, that would be a clinical diagnosis," the taller pathologist said.

As Vikram returned all the slices to the murky sweetness of the metallic bin, Dr. Mansfield thanked the pathologists for preparing the presentation. On my way out the room I stopped at one of the large gray sinks to wash my hands. I washed them with extra soap once and then once again before I wiped them dry with the rough, grainy brown paper toweling. I trailed Dr. Mansfield and interns out of the cold metallic room. It was an eerie way for me to end my relationship with Mr. Marino.

Yet another week passed, and Mr. Harvey was still in the hospital. He and his family wanted him to go home, but he continued to have abdominal pain and bowel obstruction. Every morning he pointed to his swollen, taut belly and then to the door. "Jesus Christ. Oh, Jesus Christ!" he complained.

Finally it was decided among the medical team, the oncologists, and the gastroenterologists that Mr. Harvey could go home even in his present condition. He would have increased visiting nurse services to help his family manage his complicated medical needs. Everything was set up for him at home, and an ambulance had been arranged to transport him home for the following morning.

Then, overnight, Mr. Harvey spiked a fever to 102 degrees F. A chest X ray showed he had developed pneumonia. Intravenous antibiotics were started, the ambulance was canceled, and his family was called. Mr. Harvey resigned himself to a still-longer hospital stay.

Now in the fourth and final week of my rotation at the VA, Mr. Harvey had been in the hospital as long as I had. After a few days of intravenous antibiotics, his fever broke, his respiratory situation improved, and his pneumonia was no worse by X ray. So we decided to let him go home and finish his course of antibiotics with an oral preparation. We warned his wife and daughter that he might worsen at home and require rehospitalization within the week. But they wanted so much to have him at home, and he was so eager to leave, that his family decided to take the chance. He didn't return in the few days I had left on the service.

Desperation

After nine months of intense rotations the hospital was getting to me. The enthusiasm and energy with which I approached my first months on the wards had worn thin, and I had grown tired of trying to perform for my residents and senior physicians and to fit in with my teams. I was tired of the hospital. My classmates felt the same way.

The constant transition between teams and entire disciplines was exasperating. I had to figure out a new set of personalities, a new list of unwritten expectations, and a new body of information on almost a weekly basis. Just when I established my rhythm and began to feel comfortable with the routine in one place and in one discipline, it was already time to move on to the next. I never had a sense of being fluid with the information or comfortable in the environment. Coming into the third year, I had thought the huge amount of information I was required to absorb and the intense fatigue would be the challenge of this most difficult year of medical school. But the rapid transitions, the absence of routine, and the need to generate enthusiasm constantly were far harder.

Carlos's roommate remarked, "I'm tired of constantly feeling like a puppy dog in the hospital, trailing my interns and residents. The problem is, we get tired of our current role before we have learned enough to move on."

During my outpatient month of internal medicine I bumped into Andrea at aerobics. Although we belonged to the same gym and both worked out often, it had been months since I had seen her there. As we talked after class, she said, "I don't know what it is. I'm doing adolescent medicine, which I should love. This is what I've always wanted to do. But I'm miserable! I'm exhausted all the time even though I get a full night's sleep in my own bed every night. Even now I can't find time to work out regularly. And this is the best it gets. What's wrong with me? Are you feeling like this? And the constant evaluation is driving me crazy. Then I feel frustrated that I am so dependent on good feedback. I don't generally think of myself as someone with low self-esteem, but maybe I am."

The arbitrary nature of the feedback was troubling as well. Sometimes an awkward or poor interaction with one person on the team could color an entire performance review. "At the end of my ob rotation I went to the course director for my evaluation," Masha told me one afternoon after Patient-Doctor. "This was something I wanted to do potentially, and I had felt like my month and a half on the service had gone pretty well. I walked in, all happy, kind of like 'Oh, I'll go get my eval,' and I got totally reamed. It was the first time I had ever gotten any kind of negative feedback, and this was like character-destroying stuff. I was devastated, just totally devastated. I didn't feel like any of my interactions were that bad. The thing that bugs me the most is that we can't be like normal colleagues as third years because we are always being evaluated. It's totally hierarchical."

Carlos was also devastated after an inconsistent performance review during his pediatrics rotation. He planned a career in pediatrics, so the rotation and the grade were important to him. At the midpoint evaluation the course directors were effusive in their praise of his performance. They told him what an excellent job he was doing academically and how much his team and patients enjoyed working with him. They encouraged him to pursue a career in pediatrics and suggested that he might even come to that hospital for his residency. But he had some difficulty with the intern he worked with during the last two weeks of the rotation. While he thought they had a good working relationship, he knew they hadn't clicked. At the final evaluation the course directors focused on his most recent review.

"They said they had been told that I didn't take criticism well and could be condescending. They said, 'This won't affect your grade, but you should think about this over spring break.' What could I say to that? I was between a rock and a hard place. No one in my entire life has ever said that to me. But if I disagree with their criticism, then I am just proving their point. What could I do?"

Renu, still struggling after her dismal experience in her surgical rotation, decided in February to extend her medical leave. She never rejoined our Patient-Doctor group.

In April one of the senior surgical residents committed suicide. He had been the star of his program and was slated to be the chief resident the following year, a tremendous honor. I had done my surgical rotation at a different hospital and had never met this particular resident, but many of my classmates knew him well. They said he was the only surgical resident who enjoyed teaching and was nice to medical students. He was the mentor propelling several classmates.

The resident killed himself by having an intern start an IV on him, and then, later in the day, he injected himself with a deadly cocktail of drugs. The intern had thought nothing of starting the IV, since it was fairly common practice for surgeons to give themselves IV fluids for hydration when they were sick to help them perform their grueling eight-hour surgeries.

The news traveled rapidly through all the local hospitals. All that week classmates, residents, and staff physicians talked about the suicide. How tragic, they said. How tragic, when he was almost done. How tragic, when he was such a gifted surgeon. How tragic, when he was about to become the chief resident. Later we learned he was in the process of divorce, and the day he died, friends were coming to help him move out of his house and into his own apartment.

Kate was doing her surgical rotation at the same hospital and had known the resident who died. She and I talked often about the situation in the following weeks. "Everybody says it's so tragic because he was such a great surgeon," Kate said. "But if you have a miserable personal life, it doesn't matter how good a surgeon you are. And in surgery, where the work is so hard, the days are so long, and you're chronically exhausted, it's so easy to lose perspective. What surgical resident would even have the time to make a mental health appointment? You can't just pop out of surgery for an hour. It's impossible. And even if you could, in that atmosphere, you could never say that you needed time off to go to a doctor's appointment. What's tragic is that he felt so depressed, that no one noticed, and that there was no one for him to turn to. That's what's tragic."

Andrea was on the resident's team when he died. "No one said anything," Andrea said. "It was like one day he just didn't show up to work. We were standing around and waiting for him. I heard one or two people talking about it that day, and then it was like nothing had ever happened. People were working a lot harder to cover his call and other responsibilities in addition to their own, and it was like the whole thing had never happened. It was weird," she said.

Relationships Reprise

Carlos and I got engaged.

At the end of my medicine rotation and his pediatrics rotation, we spent our weeklong spring break in California. I was surprised. I knew he had been working extremely hard during his pediatrics rotation, and I had no idea when he planned the proposal. In fact, when he told me he had a ring, I didn't believe him. As it turned out, he had bought the ring well over a month earlier, before starting pediatrics.

Even before getting engaged, we had decided to go through the match, the computerized residency placement system, as a couple. Couples match linked applicants so they had to be considered together when the computer assigned residency positions. That way we were assured of being in the same area for residency. Yet I remembered being nervous before starting our clinical rotations that our frantic schedules would chip away at our connection, causing us to drift apart over the course of the year. If we could make it through third year, I had thought, then we could surely survive medicine as a couple. After returning from a wonderful trip to Greece following the boards at the end of second year, we had sat up the night before our rotations wondering why we had chosen this profession. When we enjoyed being together so much, why did we choose a career demanding long hours apart?

Our schedules for third year were fairly well coordinated. We chose our vacation months at the same times, and while we didn't share any rotations, our difficult months and lighter months coincided nicely. Since we were only medical students, none of our rotation directors cared which nights we took call as long as we put in enough hours overall to satisfy the requirements. With a certain amount of finagling, there were only a few months where our call schedules were not coordinated. Those times were difficult. There were weeks when we didn't see each other for days on end, and when we finally had a few hours together, we both were exhausted.

Sharing third year was actually very special. It was nice to come home to someone who really understood what my day was like. The hospital experience was so intense, Carlos and I often mulled over clinical dilemmas or wrenching moments into the night until we could finally put the hospital to rest. Even something as simple as sharing the same sleep-wake cycle was important. Carlos and I both needed to wake up before six and were exhausted by ten-thirty in the evening. We shared many hurried ten-minute breakfasts at the crack of dawn before rushing off to our different hospitals. Rather than force us apart as I had feared, sharing third year brought us closer.

Carlos and I would be the second in-class couple to get engaged. The first couple to get engaged broke up just a few months later. I'd heard rumors of other couples in the works, but as far as I knew, we were one of the few.

While the intense gossip network of the preclinical years dulled somewhat as we parted ways third year, news of our engagement made the rounds quickly. We arrived back in Boston from Cali-

fornia on Sunday night, far too late to call our friends. But by
the time we arrived at our Patient-Doctor session at noon on
Tuesday, almost the entire class knew. Our classmates were excited for us. Many were surprised. We delayed starting my small-
group session because everyone was eager to hear about how we
got engaged, our wedding plans, and our residency plans. .

Coincidentally, Carlos and I were scheduled for our medicine subinternship at the same hospital the month after we got engaged.
The subinternship was an advanced general medicine rotation.
We worked on the medicine wards for one month and were expected to admit more patients per night and to follow them more
independently than we had during our three-month medicine rotation. At our hospital site, four teams divided the care of approximately one hundred patients. Usually each team had one
subintern and one or two medical students doing the three-month
rotation. But a few days before our rotation started, Carlos's roommate, who had seen a copy of the coming month's team
assignments of residents and medical students, casually mentioned
to us, "Hey, I think I saw both of you listed together. I think
you're on the same team."

At first I wasn't sure how I felt about being on the same team.
In my three years of medical school, I never had a tutorial, small-
group lab, or rotation together with Carlos. Carlos was a very
conscientious and gifted student, and I feared that after all these
years he would finally learn what a terrible student I was. Would
he be ashamed of me? What if we didn't like working together?
While the many hours apart demanded by disparate rotations
were frustrating, what if the many hours required together were
suffocating?

Carlos had fewer reservations than I did. He was eager to be
on the same team from the moment he heard it might be a
possibility. So what if we didn't like working together? he said,
which he didn't think would happen. We could still be a good

couple even if we didn't like working together professionally, he said.

"Well," I said, "our team will certainly be amused when they find out."

When we arrived at nine o'clock on Monday morning and received our team assignments, we found out that we were indeed assigned to the same team.

"I hope you two are still dating," a classmate whispered to me on our way out to meet the teams. She must have been one of the very few who hadn't heard about our engagement. When I assured her we were still together, she replied, "That's good. Otherwise it would have been really awkward."

Carlos and I decided not to make a point of telling our team that we were engaged, although we didn't hide it either. We thought it would be fun to see how long it took everyone to find out. As it turned out, it took quite a while. The team was extremely overworked, and although everyone got along, it seemed that no one talked to anyone else socially. The supervising resident didn't find out until after the monthlong rotation had ended.

Despite my initial worries, by the time the first day of the rotation rolled around, I was excited to be on the same team as Carlos. Although still a little nervous, I was glad the situation had worked out the way it did. We decided I would take the first call that night, and I was already depressed about the many hours still ahead of me by the time we left to find our team. Carlos took the edge off my bad mood. I knew he felt for me.

A senior physician and an intern stood in the hallway talking, but as they saw our group of four bewildered students, the physician called out to us, "Hi. Are you the new medical students? Which of you are on team B?"

That was Carlos and myself.

We followed the physician and intern as they went in and out of the patients' rooms. We still lugged our bags because we hadn't found our lockers yet, and each abbreviated patient story formed a blur of information neither of us could piece together. It was miserable. Half an hour later the other two interns and the supervising resident joined us for formal rounds.

Paul, our supervising resident, was in the med-peds program, a four-year residency for board certification in both adult medicine and pediatrics. He was average height with light brown hair and angular features. He was soft-spoken but also assertive. Over the course of the month I became impressed by the dignity he afforded his patients in an overloaded system that didn't easily lend itself to close interpersonal relationships.

After rounds Paul caught up with us in the call room as we dropped off our bags. "You know, we usually try to be nice to each other and help each other out. Usually the on call intern brings in coffee and breakfast for the postcall person. So if you guys want to bring breakfast for each other, I think that would be nice."

I turned to Carlos, smiling. "So, are you going to bring me breakfast tomorrow?" I asked. "I think we can manage that," I said to Paul.

I grew to love having Carlos on my team. Although there was certainly the potential for us to compete with each other, that didn't become an issue for us. We often split work we couldn't finish. If I had more time than he did, I looked up labs for his patients in the computer so he could present them on rounds. If he had more time than I did, he finished a note I hadn't gotten to. If he stayed on call and I had to send a fax before I left, he did it for me so I could get home.

As medical students we were expected to do more complete patient presentations than the interns on morning rounds. While we couldn't cut them too short without being criticized, the interns were exhausted and sick of listening. It was common for them to doze through rounds. If they weren't sleeping, then they were answering pages. It wasn't unusual to run through an entire seven- to ten-minute presentation that was unheard by any person on the team. But I knew that at least Carlos always listened to my presentations, and vice versa.

We also enjoyed sharing patients. Because we liked discussing medicine, it was fun to come home and have more detailed discussions about each patient and his or her medical management. We shared the funny things that happened on rounds and complained about the ridiculous or annoying incidents. If I was upset by how I answered a question or presented a patient's history, I trusted Carlos's perspective on the incident.

When we received our grades, however, we learned Carlos had done better than I. Initially I was a little surprised and fairly disappointed. I knew that Carlos was one of the most talented members of our class. He had an unbelievable memory and could recall details of articles long after I'd forgotten I had ever even read them, and he always did better than I did on exams. As much as I knew I shouldn't and even couldn't, I measured myself according to his standards. But in our past rotations we had done similarly, and I hadn't felt a disparity in our level of functioning during our subinternship. Exam scores felt meaningless in the face of clinical grades, so I was particularly disappointed that we hadn't been seen as equals during this monthlong experience. But Carlos as usual was supportive. He viewed me as an equal in medicine, and that helped me come to terms with these grades.

During the month I met a senior resident who had done the same residency as her husband. "It worked out great for us," she told me. "They were able to give us the same call schedule, which made a huge difference. Even sharing the ten- or fifteen-minute commute every morning and evening gave us some extra protected time together. We were never on the same team, so really, while we were in the hospital, I hardly saw him. Being on the same team would have been too much, I think. We did that in medical school once, and that was a little too close for comfort. But this way it wasn't claustrophobic at all. Sometimes I would find him at the midnight meal during call nights, and we would have some time together then too. Of course, sharing the same residency, we didn't expand each other's circle of friends. But it worked out well for us. We were really happy."

I was relieved after talking to Leslie. Not that I thought Carlos and I couldn't manage internship and residency together. But I had heard so many stories of medicine driving couples apart and destroying family life. I thought back to the dual-career medical couples who had spoken at a panel discussion during second year. They seemed to have barely survived residency. I remembered the woman who had described the first years of her marriage when both she and her husband were residents as "blindingly unhappy." It was refreshing to meet someone who had thrived while sharing residency with her husband.

Despite the difficulties of sharing the intensity of residency, having a partner outside medicine has its own challenges. One intern was married to a woman working in advertising. "You know, I can come home, and we can have a great conversation about something totally nonmedical. Or on the days that I am postcall and exhausted and can't get the hospital out of my head, we have nothing to talk about," he had said.

Another intern lamented that her husband, a businessman, didn't understand her job. "He just doesn't understand why I come home so exhausted and why I can't leave work on the drop of a dime to pick up my daughter. I try to explain to him, but he just doesn't get it."

"And there's absolutely no way for him to understand unless he does it himself," the senior resident had responded. "It's impossible. There's just no way." The senior resident had been married briefly and divorced during medical school. "My classmates were trying to organize an evening review session, and I said I couldn't make it because I had to get home and make dinner for my husband. 'What are you doing here?' a classmate asked me. At that moment I knew it was the beginning of the end," she said. "I think that's one of the big reasons I am so bitter about medicine."

However, many people enjoyed the diversity gained from a significant relationship with someone outside medicine. They found a relationship permeated by medicine stultifying and overwhelming. Among my classmates there was ongoing discussion about whether it was better to marry another physician or someone who was outside medicine.

"I couldn't imagine doing this with someone else who's doing exactly the same thing," said Andrea, whose husband was a lawyer. "I can't imagine juggling two call schedules. And I think it would be suffocating. We'd never talk about anything except medicine," she said.

Carlos and I definitely spent a lot of time talking about patients and medicine. It helped that we both had significant shared interests outside medicine, but medicine was a large part of our day-to-day relationship as we shared our experiences on the wards. We found that the amount of time we spent discussing medicine was inversely correlated with our distance from Boston. Whenever we traveled, we focused significantly less on medicine. But as we returned, our conversation gradually drifted back. By the time we were a mile or two from Boston, our conversation was dominated by our upcoming schedules, concerns about our rotations, and medical articles we had been meaning to read.

Sharing medicine worked for Carlos and me, but I could understand how it might overwhelm others.

The Birdman

The Birdman was back. He had been transferred from the
intensive care unit overnight to our team on the regular
medical floor. All the residents knew him.

I thought I had misheard when they said Birdman. The supervising resident, Paul, and the interns were talking about him
in the stairwell, and Carlos and I, per protocol, trailed behind. As
soon as I got to the floor, I realized I had heard correctly. A loud
chirrup, chirrup, chirrup traveled down the hall and filled the
corridor. No one knew exactly why he was chirping. There was
no medical reason, and he could do it on demand. The nurses said
he did it when he wanted attention.

The Birdman was a quadriplegic. Only forty-one, he had been
paralyzed from the neck down in a motorcycle accident more than
seven years earlier. He was lucky and retained the ability to
breathe without a ventilator. His wife had cared for him at home

until two months before, when she put him in a chronic care facility. He had done well in his new home until two weeks prior, when his body temperature dropped from a normal of approximately ninety-eight degrees Fahrenheit to less than ninety-two. Severe spinal cord injuries impaired the regulation of blood pressure and body temperature. It was common for the core body temperature to fall in response to infection, whereas most people would spike a fever. Indeed, the Birdman was diagnosed with a blood infection that had originated in his bladder. But even with antibiotic treatment, his body temperature and blood pressure continued to fluctuate, and he was transferred to the intensive care unit for more constant monitoring.

In the ICU the Birdman's temperature still fluctuated. For unknown reasons, he stopped breathing and was given a breathing tube. He remained on the ventilator for twenty-four hours until he was able to breathe on his own again. A day later he suffered a second respiratory arrest and was on the ventilator for another couple of days. Although the Birdman had previously designated himself do not resuscitate, do not intubate, during this hospitalization, when asked whether he would accept the breathing tube for a limited amount of time, he and his wife both had agreed.

By the time he came to us on the general medical floor, his respiratory situation had stabilized, and he was breathing on his own. His temperature remained low, but he appeared to be responding to the antibiotics.

The Birdman was no longer chirping. He lay in the center of a large bed, and the deep hum of the machine circulating air through the mattress to prevent pressure ulcers dominated the room. He must have been a tall man, because his body stretched nearly the entire length of the bed. His skin was pasty white, and his limbs were curled and atrophied. He looked over at us with his bright blue eyes. His dark brown hair was short, in an all-American buzz cut. He smiled at us with his thin, sickly lips and mouthed a few words.

Paul, the supervising resident, listened to the Birdman's heart and lungs. Satisfied with what he heard, Paul told the Birdman he was doing well.

On the Birdman's fourth day on our service we were nearly at the end of morning rounds when we were paged to the ninth floor. The Birdman was having difficulty breathing. When we arrived at his room, several nurses had already gathered, and the red crash cart had been pulled over near his room. The drawers of the crash cart contained medications and medical equipment needed to resuscitate a patient in an emergency. The Birdman lay in his bed, a green-tinted oxygen mask covering his mouth and nose. His blue eyes looked over at us. Small beads of sweat had gathered on his forehead, giving his face a sickly white sheen. Paul and the three interns positioned themselves around the bed. Carlos and I stood back in a corner of the room. Our attending, who happened to be with us that morning, stood behind me.

"Are you having trouble breathing?" Paul asked him.

The Birdman nodded.

"He had a bowel movement," one of the nurses said.

Paul pulled away the sheets to reveal a mass of gooey black stool. "That's blood," one of the interns called out. "Let's guaiac this," she said.

One of the nurses pulled out a card to test the stool for guaiac, or heme, signifying the presence of blood. But she couldn't find the bottle of developer for the test and sent me to track some down.

When I returned, the Birdman was doing worse. The stool blood test was markedly positive. Despite the oxygen mask, the oxygen level in his blood was dropping. His breaths were prolonged and labored.

"We need an ABG," Paul called. The nurse handed him a syringe kit, and he started poking in the Birdman's groin for the femoral artery to test for oxygen content (arterial blood gas). After a few tries, dark red blood welled up into the syringe. It was ominously dark compared with the brighter hue oxygen should create in arterial blood.

As we watched, the Birdman continued to deteriorate, and a nurse paged the on call anesthesiologist to the room. Because they had the most experience, the anesthesiologists were often called to insert breathing tubes when patients were crashing. A few minutes later two anesthesiologists entered the already crowded room. They carried a large yellow duffel bag packed with equipment.

Just after they arrived, the arterial blood gas values came back. The test showed gravely depleted oxygen levels, even worse than we had expected. The anesthesiologists rummaged through their yellow bag for the intubation equipment and medications in preparation for inserting the breathing tube.

"You're having trouble breathing now," Paul told the Birdman. "You're not getting enough oxygen into your blood. We might have to give you a breathing tube. Do you want the tube?"

The Birdman shook his head no. He looked at Paul.

"We're worried about you. It will probably be like in the ICU. You won't need the breathing tube for more than a couple of days. Can we give you the breathing tube?" Paul asked again.

The Birdman shook his head no.

"Call his wife," Paul called to our senior physician, who had been on rounds with us. Decisions to make a patient DNI or DNR must be documented by a senior physician. The tall, lanky physician quickly strode out of the room.

"We've gotta intubate this guy," one of the anesthesiologists said. "His blood values show that we have to intubate him. He's not breathing enough on his own."

"I just want to wait a little longer and see if he'll come around by himself," Paul said. "He doesn't want the breathing tube."

"I don't know what you're waiting for. He needs to be intubated," the other anesthesiologist grumbled.

"Let's do another ABG," Paul said.

A nurse handed him another syringe kit, and Paul again poked at the Birdman's groin. He got a deep maroon sample within a few tries. It didn't look any better than the first sample.

The physician came back into the room a few minutes later. "His wife says don't intubate if he doesn't want it."

"Do you want the breathing tube?" Paul asked him again.

The Birdman shook his head no.

The anesthesiologists were getting impatient. "We can't spend our whole day up here. This guy needs to be tubed," one of them said.

Within a few minutes the blood values came back. These results were even worse, offering proof that the Birdman needed to be intubated. One of the anesthesiologists pulled out his black rubber mask attached to the black balloon bag used to squeeze oxygen into people's lungs.

"You need the tube," Paul told the Birdman again. "You're not getting any oxygen to your blood. Most likely this will only be short term. Do you want the tube?"

The Birdman looked at Paul and shook his head no.

"Do you understand what I'm asking you? You could die," Paul asked.

The Birdman nodded.

"Okay, we're going to try to do this without a tube. Let's transfer him to the ICU. We can do mask ventilation there. Then, if he decompensates further, they are well equipped to do an intubation on the floor," Paul said. Mask ventilation was noninvasive, and in the intensive care unit the staff is accustomed to placing breathing tubes.

"But this guy needs to be intubated. We can't transfer him this way. He's too unstable," the anesthesiologist protested.

"We can bag him until he gets to the ICU. He doesn't want this breathing tube, and we need to respect his wishes," Paul said. They could continue to use the black balloon bag to force oxygen into his lungs.

In the end Paul prevailed. The Birdman was wheeled out in his big air bed, and the anesthesiologist walked along one side, rhythmically squeezing the black balloon bag, while Paul walked along the other.

I had a sour lump in my throat as I watched the Birdman shake his head and refuse the breathing tube over and over again. Often when patients were in the midst of an acute crisis, they wanted to live despite any previous arrangements they might have made regarding code status. Yet the Birdman did not waver in his refusal. I watched as he chose death. It was painful to let him make that choice, knowing full well that the breathing tube could easily get him over this likely time-limited crisis. At the same time I was impressed by the tremendous sense of dignity Paul brought to the crisis. He gave the Birdman the opportunity to refuse the tube and retain his autonomy even to the death. Paul was able to disregard the anesthesiologists agitating to intubate the Birdman. For the first time in my year in the hospital I felt as though we had done something right. I felt that we, through Paul's lead, had been humane.

It was difficult as hospital caregivers to refuse to use the potentially lifesaving technology and allow patients simple, noninvasive deaths. It was difficult to let our patients go without a fight. Death with dignity was a rare occurrence in the hospital.

Later that morning, a fiber-optic camera demonstrated a bleeding ulcer in the Birdman's stomach responsible for the bloody bowel movement, but he had not lost a significant amount of blood. The Birdman recovered in the intensive care unit, and he never required the breathing tube. A few days later he was ready to go back to the regular floor again.

Carlos and I bumped into Paul and our senior physician eating lunch a few days later. They were talking about the Birdman when we got there.

"You know, Paul, I'm so impressed that you were able to give him the chance to refuse the tube and then supported his decision. Especially with everyone else in the room fighting to get him tubed," I said.

"Well, in the ICU they said we should have just tubed him," Paul said. "Even though he and his wife had decided that he should be DNR/DNI before this admission, they had already flip-flopped twice during this hospitalization on whether he should be intubated or not. That demonstrated ambivalence on both their parts as to how ready they were to let him go. As long as the intubation was anticipated to be short term, he should have been tubed. The family needs a cooling-off period to think about their situation and reassess his code status."

I was disappointed. For the first time I thought we had done something pure and right, and then I learned that maybe we had actually made the wrong decision. I felt disillusioned. Maybe it was never possible to have death with dignity when one was faced with the ability to prolong life. But at the same time, the reasoning of the ICU physicians made sense to me. We needed to give people the opportunity to choose life. Resuscitation decisions were so complicated. How could we ever get them right? The public often criticized the medical profession for needlessly prolonging life. But how could we stand by and let someone die if we weren't sure he or she was ready to go?

A few days later we were seeing one of our patients when there was a code call from the dialysis wing. A patient was dying. We all rushed over, and by the time we got there, people were already clustered around the patient, giving shots of epinephrine and chest compressions in an attempt to revive his failed heart. Suddenly one of the nurses came rushing in, waving her hands. "Stop! He's a no code! Stop! He's DNR!"

We had to act because there wasn't time to think. The time lost checking for a code status or consulting with a family member could cost the patient her life. And if it's not only we but the patient herself or, even worse, the patient's family who were not ready to say good-bye, it was impossible to know how to manage the situation. We had to try to save everybody in the hope that we would someday save the one person who really didn't want to die.

This realization made me feel bad about my role as a medical caregiver. After working with hospice, I strongly believed in al-

lowing patients to die simply, without interventions. I treasured
discussing the options with the patients and their families, listening as they decided when they were ready to let go. But as a
doctor I will be forced into a role where there is no time to be
thoughtful. I will never be sure if I made the right decision and
will have to take small comfort in knowing I have made the only
care choice possible in the situation.

Psychiatry

The inpatient adult psychiatry unit was on the top floor of the farthest building to the right in the interconnected maze of buildings composing the hospital. The unit was a locked ward. But it hadn't always been this way. Just a few years before, it was an open unit; patients were there voluntarily and could leave at any time. But with managed care, insurance companies decided that if a patient didn't require a locked door for safety, then she did not need hospitalization. The open unit was forced to install gray metal locks on all its doors to protect its funding. Now all the psychiatric inpatient facilities in this hospital, like most other wards in the rest of the state, were locked. A photocopied sign taped to the front door of the pediatric unit warned SPLIT RISK! in goofy black lettering over a swarm of smiling cartoon children.

The elevator to the fourth floor opened into a small foyer with thick glass windows looking into the adult unit. A pink metal door marked the boundary between inside and out. No other doors or corridors led off the pink foyer. Once the steel gray doors of the elevator shut behind me, there was no alternative to the psychiatric unit. Through the windows I looked into the hallway and the nurses' station. A few patients paced up and down the hall; one stopped for a glass of ginger ale set out on a small table in front of the nurses' station. A tall woman with brown curly hair wearing a retro smock top with bell-bottom jeans stood in front of the station. She carried a clipboard under one arm, and an electric pink coiled plastic key chain dangling a single silver key stretched around her wrist. I rang the doorbell, and although I couldn't hear the bell through the thick glass windows, the woman smiled at me through the window, reached for her silver key, and unlocked the door. "Can I help you?" she asked.

The unit itself was small. The ward was laid out on an L-shaped hallway, with the entrance directly opposite the nurses' station situated in the central angle. The bedrooms, the dayroom, the five small meeting rooms, and the two large-group meeting rooms branched off a narrow hallway that took less than one minute to traverse end to end. The eighteen patients on the unit at any given time were confined to this space; they had to earn privileges to be released for even a short time. All the wooden doors off the hallway had gray metal locks, and except for the bedrooms, each had a narrow glass window. Every five minutes a staff person peeked through these glass windows to keep tabs on the occupants within.

Each bedroom had a cardboard sign posted outside identifying its occupants. CAROLYN S.; TOM D., JIM B., MICHEL H.; JESSICA B., NATALIE M. The sparse rooms held from one to three white cots, usually disheveled with white cotton hospital blankets haphazardly strewn across them. A single bay of wooden lockers, also with cardboard nametags, lined one corner of each room. Otherwise the rooms were bare, strangely devoid of photographs, cards, or other knickknacks to offer a sense of personality to the generic space.

The dayroom was the only room on the ward unlocked and

available to the patients at any time. The dayroom had a long, narrow table with wooden chairs stretching nearly the length of the room. Softer lilac vinyl chairs rimmed the periphery. A bay of windows took up the longest wall. A large TV in the far corner was always playing, and a worn wooden upright piano sat tucked in the corner on the right. Patients gathered in this room to play cards and Connect Four, to socialize, and to watch TV. The dayroom doubled as the dining room at mealtimes. At eight, twelve, and five, the doorbell rang, and a food services employee wheeled in a tall metal cart stacked with cafeteria trays. One of the paranoid patients caused quite an uproar by taking other patients' trays; she thought hers was poisoned. After a few days her medication took effect, and the situation resolved itself.

The "quiet room" was in the farthest corner of the hallway. Sometimes patients chose this space voluntarily when they couldn't otherwise control their acting out, but they were also taken there forcibly when they required physical restraint. More than once during my month here, blue-uniformed security guards were called to contain a patient and escort him or her to the quiet room. Although this room was the topic of much discussion, I never actually saw its confines. It was strictly for the patients.

After almost a year of rotations I was well accustomed to the patient-clinician dynamic. While surgery, medicine, obstetrics, and pediatrics were vastly different, the patient routine was relatively the same. We reviewed patient issues each morning on rounds and reassessed the care plans. While each specialty valued a slightly different set of information, I was familiar with the daily litany of vital signs, events, and interventions common to all. But psychiatry was something else altogether.

In morning report the nurses gave the rest of the medical staff a rundown on how the patients had fared over the last twenty-four hours. "Edward was brighter and more visible on the unit yesterday. He spent the better part of the afternoon in the day-room," or "Randall was seen crawling up the back staircase and dragging his wheelchair behind him yesterday afternoon when he was returning from pass. We would appreciate it if you would address this issue with him." These reports also included a list of the patients who attended the previous day's group therapy sessions. "Natalie, Jessica, Brian, and Edward attended occupational therapy and made T-shirts. Jan drew a black rose on hers but would not comment on what that meant to her. Natalie completed a T-shirt for her daughter and talked about how much she missed her children. The session was lighthearted, and a good time was had by all." After the concreteness of vital signs, lab values, and test results, it took me nearly half the month to get used to listening to these daily reports and to understand how all this information that seemed trivial, silly almost, became meaningful and significant.

During rounds, instead of talking in the hall and traveling to each patient's room as we did on my other rotations, the patients were requested to come into the group therapy room and participate in a short interview with their primary caregiver in front of the team. At these meetings the patients were questioned about their symptoms, any side effects to the medications, and their goals for the day. The patients were also expected to bring their requests for privileges, "privs."

Greta was a twenty-six-year-old Austrian woman in the United States illegally. She had frizzy brown hair carefully braided on her good days, greasy and tied back in a loose ponytail on her worst. Her freckled arms had pale pink scars piled up on both her wrists. As a child she had suffered terrible physical and emotional abuse. Greta had bright blue eyes that suddenly clouded over and went blank as she lost touch with her body, with her surroundings. "Greta! Greta!" the psychology intern yelled, kicking the chair. "Greta!" Then, as suddenly as she left us, Greta returned, blinking, and continued to answer the question just where she had left off.

"Can I have grounds today?" she routinely asked us in her accented but perfect English.

Grounds meant that patients were allowed to go alone anywhere on the hospital property for periods of fifteen minutes at a time, which could be increased as they proved they could manage appropriately. Greta had a long history of self-mutilation. Even on that stay she had transported razor blades back to the unit concealed in her vagina.

"What would you do if you got overwhelmed outside, if you felt afraid?" the psychologist asked.

"I would come back."

"What would you do if you found a piece of glass on the ground?"

Greta thought for a minute, then looked straight at the psychology intern with her dull blue eyes. "I would cut myself."

"Okay. Thanks for your honesty. We'll have to talk about it, and I'll tell you the answer later."

We'll have to talk about it. That answer often bothered patients. "But," they said as we shepherded them out of the room, "can't you tell me the answer? Can't I have grounds today?" They wanted to know now, right away. But I think they were also troubled that twelve people would sit and, behind their backs, pass judgment on them and their ability to cope. Perhaps I projected, unconsciously imposing my concerns on the patients and making false assumptions about their worries on the basis of my own feelings. But if this concept of judgment, of micromanagement, didn't trouble our patients, it certainly disturbed me. This system of exerting tight control over patients has been described as taking over the responsibility of the superegos. Patients who came into a locked ward felt out of control, lost. Their superegos, normally functioning to provide structure so they could behave appropriately, had weakened. The psychiatry staff gave them a sense of external structure by which, in addition to psychotherapy and medications, they could regroup and reorganize their lives. The rationale behind physical restraints also stemmed from this idea: A patient acutely out of control of his or her actions yearned for a source of external constraint to prevent the destructive behavior. Nonetheless, no matter how convinced I was that someone wanted me to take control of the most minuscule details of his life, this kind of control did not sit well with me. I felt awkward each time a patient left the room with the question unanswered, as if I were complicit in an act of betrayal.

In psychiatry, like pediatrics, most of the clinicians chose not to wear the white coat to put the patients more at ease. But I think discarding the white coat was as much for the clinicians as for the patients. On the psychiatric ward the power discrepancy between patient and doctor was more pronounced than in any of the other medical relationships I had yet experienced. With our ability to give and withhold privileges and prescribe drugs that alter the function of the mind, it felt as though we controlled every detail of our patients' lives while they lived in the locked ward. There were many legal safeguards to protect the autonomy of the patients and to give us a more comfortable framework to work within. Nevertheless the power rested uncomfortably on the caregivers' shoulders, and the bleached stark white coat would be a painful visual reminder of this unease.

Jessica

Jessica was the youngest patient on the floor. She was only nineteen, and this was already her second hospitalization. She had been admitted to another local hospital seven months before for an acute psychotic break when she was paranoid and hallucinating. Her psychiatrists thought she might be schizophrenic. But afterward, on medication, she did much better, returning to her prior level of function.

Jessica and her mother both hated the side effects of the medications, however. Her mother complained that they turned Jessica into a zombie. While the medicines blunted the paranoia and hallucinations, they dulled her personality as well. Because Jessica was doing so well, her psychiatrist agreed to stop the medications. It had been only Jessica's first psychotic break, and her psychiatrist hoped that she didn't really have schizophrenia, that her psychosis had an alternate explanation. Schizophrenia is difficult to diagnose, requiring many months of observation, and it carries a poor prognosis. Despite treatment, most people never fully recover after their first breaks, which usually occur between the ages of seventeen and twenty-five.

Initially Jessica continued to do well off her medications. She worked at a drugstore and enrolled in courses at a local community college. She spent a lot of time studying for her midterms, determined to do well. Yet the exams weren't what she expected, and she did poorly. Her mother said this failure sparked a depression. In the few weeks prior to this admission, Jessica became more withdrawn and spent more time by herself in her room. A few days before her current admission Jessica came downstairs and acted as though she were catching and then throwing a ball. "What are you doing?" her mother asked her. Jessica informed her she had spent the weekend in New York and had kissed a boy, although her mother knew she had spent the days in her room. Over the next few days Jessica became even more mentally confused and disoriented. She became concerned that the people in the apartment upstairs were torturing her mother. On Monday her mother took Jessica to the psychiatrist, who sent her straight to the emergency room.

Jessica was "pink slipped" into the unit. Because the unit was locked, patients had to sign forms saying they agreed to the treatment conditions and wished to be patients in the psychiatric unit. But, like many of the other acutely ill patients on the floor, Jessica didn't think she required hospitalization. The emergency staff always tried to convince unwilling patients that they needed treatment and encouraged them to sign the form voluntarily committing them to a locked ward. When patients would not sign, the emergency department staff resorted to a Section 12. More commonly known as the pink slip, this provided physicians with the legal means to force patients into the hospital. There were strict criteria for involuntary commitment to prevent abuse and to protect the patients' civil rights. If the physician in the emergency room believed a patient would harm herself, would hurt someone else, or was too impaired to take care of herself, he could legally commit her to the hospital for ten days. After ten days, if the patient had not been discharged or signed a voluntary commitment form, the physicians went to court for permission to continue to retain her on the unit against her will.

In Jessica's case she heard voices telling her to kill herself, so the emergency room physician worried about her risk for suicide. Jessica was tall and thin. Her long brown hair hung in greasy strands, and her golden skin glistened with an oily sheen. She seemed agitated when she arrived on the floor. Her large brown eyes froze for long moments at a time as she listened to the voices in her head. She wore a dark pink chenille short-sleeved sweater and a pair of old jeans. She spent much of her first day on the ward in her bedroom, refusing to come out.

On the ward Jessica refused the medications that would release her from her paranoia and from the blur of voices in her head. She wouldn't eat, and she wouldn't drink. It fell to me to convince her to take her medication.

While the pink slip could legally commit patients to the unit involuntarily, they were still considered competent to make medication choices. The psychiatrists had no power to force patients to take medications without going to court, unless they were in imminent risk of harming themselves or other people. The medication acted as "chemical restraint." Taking them into a locked ward where they could be monitored for safety and protection was drastic but not particularly invasive. The power implications of forcing a mind-altering chemical substance into another person, on the other hand, were troubling.

Yet to make an informed, competent choice, patients had to understand the nature of the medication, the possible risks and side effects, and the result of not taking it. I thought it highly unlikely that many of the pink-slipped patients had a sophisticated understanding of the treatment options. Yet patients choosing to take medication were given the drugs, whether or not they were truly competent to make that choice. Only patients who refused the medication were formally evaluated for competence and taken to court. It seemed odd to me that a physician could commit a patient to the hospital involuntarily, formally asserting that this person was unable to care for himself, and still believe that the same patient could make a reasonable treatment choice. Perhaps it was possible to understand this by one's assuming that the competent choice was not over the medication itself but over giving up control to the physicians and other caregivers on the unit.

I found Jessica in her bedroom. She lay underneath the rumpled white covers but was not asleep. I asked her if she would talk to me for five or ten minutes; she smiled and agreed. I offered her a few minutes to get up and collect herself, but she immediately pushed away the covers to reveal herself fully dressed in her pink sweater, jeans, and the green Styrofoam hospital slippers. She quickly stood up, and she followed me out as we looked for a quiet place.

The music room was unlocked. I never understood why it was called the music room because there were no instruments in it, or even a tape recorder. It was just a small meeting room with a few soft vinyl chairs, a desk, and a small table. The room was very narrow; Jessica chose the chair under the window. The circular table was to the left of her chair, and my chair was diagonal to hers, against the side wall. I was a little concerned about this arrangement because I knew paranoid patients often found a face-to-face conversation confrontational and intimidating. They preferred to sit next to the interviewer, both looking ahead with minimal eye contact. Jessica sat perched on the edge of her chair, her legs carefully arranged together. She took off the slippers and placed them carefully on the table next to her. Now she was slightly hunched over, deeply engrossed in observing her bare feet, which she tapped rhythmically, first one and then the other.

"Jessica, how are you doing?" I knew this might not be the best way to start the conversation. Paranoid patients often find the traditional question-answer interview format threatening. But I couldn't think of anything else to say.

Jessica looked up as though about to say something, but then her brown eyes widened slightly and adopted a somewhat glazed stare. She looked directly at me, staring straight into my eyes. "Jessica? Jessica?" I asked. She kept on staring.

After what felt like minutes, she blinked and looked quickly around the room. "What did you ask me?" she said.

"Jessica, what happened just then? Were you hearing voices?" She stared at me blankly again. I started to feel nervous. I had no idea what was going on in her head. Finally she nodded yes.

"What are they telling you?" I asked her.

She again stared, directly into my eyes. I looked away, unsure whether I should meet her gaze or avoid it. She bent over to stare at her tapping feet again. Then she looked back at me. She suddenly stood up, pausing slightly at half standing and then rising to full posture. She stared out the narrow glass window in the door.

"Are you feeling uncomfortable? Would you like to leave?" I asked her.

Jessica looked back at me. "Yes."

"Okay, we can leave." I was relieved. She made me feel creepy. I felt my heart beating faster, and my underarms and palms were damp. I stood up to go to the door.

"Sit down," Jessica said.

"Do you want to stay?"

Jessica nodded her head. After I sat down, she sat down as well. I felt caged in this small room. She wouldn't let me leave, and I had no idea what she would do. I looked at my watch. We had already been in the room fifteen minutes, far longer than the five I had planned on. Because the room was unlocked, I hadn't told anyone where we were going. My back was to the door, and I couldn't see if anyone was making the usual five-minute checks. I was scared.

"I'm afraid I won't have the chance to talk with you like this again," she said.

"Don't worry," I told her. "I'm here to help take care of you, and I'll be here to talk with you every day."

Jessica's eyes widened again as I talked, and she glazed over into her direct stare.

"Are you hearing voices?" I asked again.

After a long delay she blinked and then nodded.

"What are they telling you?" I asked again.

Jessica didn't answer. After a long pause she turned to me. "Are you homesick?" she asked.

I was sure she was talking about herself, and I wasn't sure how to answer the question. "I'm not feeling homesick right now, but maybe you are. It must be frightening to come to this new place all by yourself."

Jessica's eyes glazed over, and she went back to staring at me. "Yes," she said after another long pause.

I felt more and more uncomfortable. I was sweaty, and I heard my blood pulsing in my ears in the quiet room. I was desperate to end this interview. "Do these voices tell you to hurt yourself?" I asked her.

She shook her head no.

"Do they tell you to hurt someone else?"

Jessica nodded. Then her eyes went blank again, and she stared straight into my eyes. "I'm hungry," Jessica said.

"Do you want to go out and get some food?" I asked her.

Jessica shook her head no.

I was desperate to get out of that little room. I felt claustro-phobic. I knew that these patients usually aren't organized enough mentally to plan an attack, but I looked her over anyway. She was definitely taller and bigger than I was, but I thought I could probably hold her off.

"Jessica, I think we've talked enough for now. But I'm con-cerned about these voices. You seem frightened. We can give you some medicine, and it will take the voices away, and then you'll be less frightened. What do you think?"

Jessica looked at me for a minute. "Okay," she said.

"So I think we should end our talk. We both should get up and go out of this room. We can go get some medication for you outside. Then I can come back and check in with you later and make sure you're doing okay."

"Okay." Jessica remained seated.

I stood up. "Let's go now," I said.

Jessica still sat.

"Will it help if I open the door and go out first?"

"Yes."

Now she stood up, but she stayed backed against her chair. I took the few steps to the door and opened it. Jessica picked up her green slippers and took a few steps toward me. I went out the door and held it for her. She took the open door and then followed me out of the room. As I stepped into the hallway with all the people walking around us, I immediately felt a flood of relief. I felt in control again. I felt safe.

"Come on, Jessica, let's go get some medicine for you."

"I can't hear you," she said to me.

"Are the voices getting in the way?"

"I still can't hear you," she said.

"Are the voices getting in the way?" I yelled.

"Wait a minute." She put her hands over her ears.

After a long pause I asked, "Are you okay now? Can you hear me now?"

Jessica nodded.

"You see, this is why I'm so worried about you. I think these voices are troubling you a lot. The medicine will take them away. That's why it's so important to take the medicine."

Jessica followed me the few more feet to the nurses' station.

Bill was the nurse in charge that afternoon, and I liked him a lot. He was young, with a thick Boston accent. He never patronized or pitied the patients but approached them with respect and empathy. I was relieved to have him help me with Jessica.

"Bill, Jessica has decided that she would like to take her medication," I told him.

"Great! Is that true, Jessica? Do you want to take your medication?" he asked her.

Jessica nodded.

Bill went into the back medicine room and dispensed one brown tablet of antipsychotic medication into a small plastic cup. He returned with the plastic cup and handed it to Jessica.

"Here you go, Jessica. This medication will help you feel better."

Jessica brought the plastic cup to her mouth and then took it away again without taking the pill. It sat at the bottom of the plastic cup.

"Jessica, remember how the voices were so bad that you couldn't even hear me talking a few minutes ago? The medicine will help make those voices go away. It's important for you to take this. I'm worried about you," I told her.

Jessica brought the plastic cup to her mouth again and paused with the plastic cup tipped forward and the little brown pill touching her bottom lip. She looked at me again.

"Jessica, the medicine will help you," I told her again.

She tipped the cup a little farther forward, and the small brown pill fell into her open mouth. She turned to look at me again.

"She didn't swallow it," Bill said to me. "Jessica, do you want a glass of water?"

Jessica shook her head no.

"Jessica, we can't play games here. You have to either swallow the pill or spit it out," Bill told her.

"Jessica, the medicine is good for you. It will help with the voices," I told her again.

Jessica continued to hold the pill under her tongue.

"Jessica, you have to decide now. Either swallow the pill or spit it out," Bill said. "I think it will help you, but it's okay if you don't want to take it. But now you have to either take the pill or spit it out." The unit had had trouble in the past with patients hoarding pills by hiding them in their cheeks and then taking an overdose later or distributing them to other patients.

"Jessica, the medicine will make the voices go away. The pill is safe. You can take it," I told her.

Jessica turned and looked at me. She gave a swallow.

Bill turned to me. "I don't think she actually swallowed it. It's still under her tongue."

I was surprised. I had assumed that she swallowed the pill with her gulp.

"Jessica, are you going to take that pill or not? We can't play games here. Are you going to swallow the pill?"

Jessica shook her head no. Bill held up the garbage pail, and Jessica leaned over and spit the little brown pill out. She turned and walked off down the hall.

"Shoot, I thought she was really going to take it. She really needs that medication," I said to Bill.

"I know. It's too bad. But she was close. She was thinking about it. Maybe she'll do it later."

After Jessica walked off and I had a few minutes to think about our interaction, I felt silly for being so scared. The prolonged staring directly into my eyes was such an aggressive posture that I felt threatened by her. I'm sure she didn't actually choose to be aggressive but was paralyzed by the tumult of voices in her head. Yet I couldn't help responding to her. It took so much energy just to sit in that room with her. I couldn't imagine sitting with this type of patient on a daily basis. It was too painful.

Unfortunately Jessica didn't take her medication as we hoped. She didn't take it all that day or the next. Her brown hair grew greasier and stringier. She developed two big pimples, one on her forehead, the other on her cheek. She wore the same pink sweater and torn jeans. She remained paranoid, refusing to eat or drink any food on the unit.

By Thursday Jessica showed signs of dehydration. Her heart rate increased dramatically, and her eyes and mouth, normally liquid and moist, were parched. The staff became concerned about her continued refusal of fluids. Dehydration could be very serious because of the electrolyte imbalances it caused.

"I think we're going to have to give her IM meds today if she doesn't agree to drink something and continues to refuse her meds," my resident, Neil, told me. He wanted to give her an intramuscular (IM) shot of medicine, essentially forcing her to take the medicine, if she continued to refuse her pills.

Neil and I discussed the situation with one of the senior psychiatrists. "Well," he said, "if you think her mental status is so impaired that she is at risk of causing harm to herself by refusing liquids and becoming dehydrated, then I think it is perfectly reasonable to give her IM meds today. I would go ahead and give her the shot."

Neil turned to me. "Why don't you do the talking? You go get the charge nurse—I think it's Kevin today—and I'll go draw up the meds. Then, when we go in, you can be in charge. Okay?"

"Okay," I said, and left in search of Kevin. I was nervous. I had never seen anyone forced to take meds, and I wasn't sure what it would be like. What was I supposed to say? How was I supposed to act?

When I found Kevin, he was pleased we had decided to give her the shot. "Oh, good. She really needs that medication. She'll be much better off for it. I'll go collect a few other people so we can have a show of force," he said.

"What's a 'show of force'?" I asked.

"Well, we like to collect a few people to go into the room with the patient. That way the patient realizes we mean business and is more likely to consent to the meds. Then, if they don't consent, the extra people will help restrain the patient. We basically use the people to hold the patients down on the bed, the same way you'd restrain a kid."

Kevin was from Ireland, and he spoke with a thick lilt. I liked listening to him read off morning report, but I also thought he was an excellent nurse. If Kevin thought Jessica needed her meds immediately, I trusted his opinion.

Nonetheless I wasn't terribly comfortable with the method. The "show of force" bothered me. I felt I was about to bully Jessica into taking her medication. While I fully agreed she needed it, I felt ugly about my involvement in this process.

A group of five of us collected at the nurses' station, and a nurse handed me the pills in a plastic cup and passed Neil a syringe filled with clear fluid. The pills were still packaged in their silver blister packs because no one really thought she would take the pill.

We trooped down the hall to Jessica's room. She lay in bed, but she wasn't asleep. The others ringed her bed as I stood nearest Jessica at the head of the bed.

"Hi, Jessica," I said. "I'm sorry if we're bothering you."

Jessica stared straight at me.

"Jessica, we're all worried about you. I'm concerned about these voices in your head. You seem so frightened to me. We want to give you some medicine to help you and make the voices go away. This medicine is very important for you. Will you take the medicine now?"

Jessica nodded her head.

"Really? You want to take the pills?"

Jessica nodded again.

"Okay. Look, I have them right here," I said. I got the blister packs and began to push one of the pills through the foil packaging. My hands shook slightly. "You're going to take the pills?"

Jessica shook her head no.

"No?" I asked.

Jessica shook her head again.

"I know you've been feeling ambivalent about taking this medicine. It must be scary for you to take these pills. I promise they're safe. But we're very worried about you, and I promise the medicine will help. Because we're so worried and because we're so convinced that the medicine will help, you have to take the medicine now. You don't have a choice. You can have the medicine either as a shot or as a pill. It's up to you. But you have to choose one or the other," I told her.

"Okay. The shot," she said.

"You'd prefer a shot?" I asked her. "That's fine."

"Yes, shot," she said again.

Neil started to come forward with the shot.

"No, neither," Jessica said.

"We're worried about you. You have to choose one or the other. You have to take the medicine now. We wouldn't do this if we weren't so concerned about you," I said.

"Neither," Jessica said again.

"Okay, then. We'll have to give you the shot," I told her.

The nurse stepped forward, and Neil handed her the syringe. "Jessica, honey, lie down for me and roll over onto your belly."

The nurse pulled off the covers. Jessica still wore the pink sweater, but she had traded her jeans for a pair of navy blue sweatpants. Jessica obediently shifted herself down in the bed and rolled over. The nurse pulled down her sweatpants and underwear, wiped her right buttock with an alcohol pad, and quickly inserted the syringe, depressed the plunger, and pulled it back out.

"Thanks," Jessica said, her voice slightly muffled by the pillow and blankets.

As we walked back to the nurses' station, I talked with Neil for a few minutes. I was relieved that the task was done and that it had gone so smoothly. "You know, it's really interesting to me that she thanked us after we gave her the shot. I think she was paralyzed by indecision," I told Neil. "I think the choice of whether or not she should take the medicine overwhelmed her, and she needed us to take that choice away from her."

Jessica slept for the rest of the afternoon. At four-thirty, she finally woke up. It was a gray day out, and the room had already assumed the dusk of early evening. Since she still hadn't drunk anything, Neil and I went in to see if we could persuade her to take some water. Jessica sat up in bed. A lilac-colored plastic pitcher of water, a half-filled bottle of ginger ale, and a stack of paper cups rested on the small table next to her bed.

"Hi," she said to us as we walked into the room.

"How are you feeling?" Neil asked her.

"Much better," she said.

"Are you still hearing the voices?" he asked.

"Not too much. Why don't you sit down?" Jessica motioned to the foot of her bed and a chair in the corner of the room.

"Thanks," Neil said as he sat down on the bed. I pulled up the chair. "You know," he said, "it makes me really happy to see you feeling so much better. I don't know if you remember, but you were feeling pretty bad there for a couple of days. You were hearing a lot of voices."

Jessica smiled. "No, I don't really remember. You want something to drink?" She offered us her pitcher of water.

We both said no. "You know, even though you're feeling better, we're kind of concerned because you haven't been drinking too much over the past few days," Neil said. "We were hoping that maybe you would drink something for us. Maybe we could pour you a glass of water. Do you think you could drink that?"

Jessica shrugged her shoulders.

"Are you worried about the food here?" Neil asked her. "Are you concerned that it might be poisoned?"

Jessica nodded.

"Well, how about if we all drink together? All three of us. How would that be?" Neil asked.

"Okay." Jessica smiled.

"What'll it be? Water or ginger ale?"

Jessica considered for a minute. "Ginger ale."

"Okay, will you drink out of this bottle, or should I go get a fresh one?"

"A new one." Jessica smiled.

In a few minutes Neil returned with a new bottle of ginger ale. He opened the bottle while I passed out paper cups. Neil sat on the edge of the bed, and Jessica sat Indian style at the head of the bed. Her hair was stringy, and the legs of her sweatpants were pushed up to her knees. She wrapped the white cotton blanket around her shoulders. The difference from the morning was striking. Although she was still unkempt, her eyes had lost the glazed stare. She was much more alert and communicative. I sat on the chair, and we formed a little circle together. It was still dim in the room. None of us had thought to turn on the lights.

"We'll do shots. How about that? We'll do ginger ale shots,"

Neil said. Jessica giggled. He filled Jessica's cup all the way and ours halfway. "Okay," Neil said. "We need a toast. How about it, Jessica, are you going to make a toast?"

Jessica laughed again. "To love, happiness, and a good life," she said. We clinked our paper cups in the dim room and drank the ginger ale. Jessica gulped hers down.

"Your turn," she said to Neil.

Neil refilled our glasses. "To love," Neil said. We clinked our glasses again.

When Neil refilled our glasses, it was my turn to offer a toast. "To good thoughts," I said.

"To good thoughts," Jessica and Neil echoed. We clinked our cups a third time and downed the ginger ale. We sat in the calm dimness of the room for a few minutes longer, none of us saying anything, not wanting to break the moment. I felt at peace. We had created a happy space for the three of us, doing ginger ale shots and making toasts in a hospital room. We had fun. We connected.

"Okay, so now you have a safe bottle of ginger ale," Neil finally said, breaking the silence. "You need to drink as much as you can over the next few hours." He screwed the top back on the bottle and placed it carefully on the table next to her. "This is your bottle. And you can get new ones from the kitchen whenever you want."

The next day Jessica continued to take her medication. She wanted it. As the medicine began to take effect, she spent more time with other patients on the floor. A couple of times I saw her in the dayroom playing cards, and at lunchtime I saw her eat a package of cookies at the long table with all the other patients. She talked and laughed with them.

I was a few minutes late to the weekly staff meeting later that afternoon, and when I walked into the dayroom that afternoon, everyone was engrossed in a discussion of our decision to give Jessica the shot of medicine.

"Really, if the concern was dehydration, then the appropriate treatment would have been placing an IV to give her intravenous fluids, since she wouldn't take them by mouth. The mental status was really a side issue," one of the residents said.

"Maybe you should have gone to court," one of the social workers on the unit said. "She wasn't being violent; she was just sitting quietly in her room, not disturbing any of the other people on the floor."

A patient opened the door to the conference room. She wore bright fuchsia lipstick that strayed beyond the outline of her lips and generous blue eye shadow. She had headphones on to hold back her tousled brown hair. "Chris, I need to talk to you. You said I could go out, but they—"

"Not now," the senior psychiatrist told her. "We're having a meeting. He'll come see you later."

"But he said—" One of the interns got up and pushed her out of the room as he forced the door closed. Deferred but determined, the patient went to the window looking into the dayroom and stared at us for the rest of the meeting.

I was surprised to hear how negative everyone felt about our decision. We had been so convinced it was the right thing to do. I hadn't really considered giving IV fluids. As I had thought about it that afternoon, the problem was Jessica's paranoid state of mind, and the dehydration a symptom of the primary problem. If we fixed the primary problem, then the symptom of dehydration would resolve as well. We framed the problem that day on the basis of the mental status, not the dehydration itself, and the medication had seemed to be a reasonable intervention. I hadn't thought we were abusing our privilege. Yet now, listening to these perspectives, I worried that perhaps we had rushed into the solution; maybe we had infringed on her right to refuse treatment. It seemed counterintuitive to watch Jessica's marked improvement in her ability to think coherently and her diminished anxiety and conclude we had done the wrong thing. But maybe we had. Maybe *I* had.

Usually I thought carefully about what I did, particularly when inflicting procedures or medications. I thought I was biased against invasive interventions generally. Yet somehow this time a different perspective had slipped me by.

Afterward Neil discussed the meeting with me. "You know, it was weird. At the beginning, the part you missed, we just mentioned that this had happened, and everybody agreed immediately. They were like, 'Of course you should have done it. It was a no-brainer.' But then, by the end, as all these issues came up, it felt like everyone pretty much thought it was the wrong decision," he said. "But you know, it just seemed like the right thing."

As I walked away to catch up on some work, I caught sight of Jessica walking down the hall. We might not have made the most ethical decision, but if I had been in her position, knowing what I knew now, I would have wanted my caregivers to get those medicines into me any way they could. I couldn't imagine living in the hell her mind had created.

As Jessica continued to take the medication, I was amazed by how rapidly she recovered. I thought she was doing great on Friday when I left, but I arrived on Monday to find a totally new person. Jessica prepared a written statement for morning rounds describing how she had enjoyed her weekend passes out of the hospital.

Before she left the room, Neil asked, "Jessica, do you remember me or Ellen from last week?"

Jessica looked back at me. "I kind of remember you, but I don't really remember her."

"Do you remember much of last week? You might not because you were having a really tough time."

Jessica shook her head and smiled as if a little embarrassed. "No, not really."

"What do you remember?"

"Well, I remember getting a shot. But not much else."

During her last few days in the hospital, I asked Jessica several times to tell me about the experience of getting her shot of medication.

"I don't know. Like, what do you mean?" she said when I asked her about it.

"How did you feel? Was it scary? Or painful? Were you relieved?"

"I don't know. It was okay." She was never able to explain how she felt.

Did we do the right thing for Jessica? I knew she needed that medication. Maybe we should have gone to court to ensure that a fair decision was made. But despite all arguments to the contrary, looking at her before she went home, I truly believed we had done the right thing.

During that month I was never quite comfortable with my role as a psychiatric caregiver. I left with a much deeper respect for psychiatry, however. Other physicians joked about the cushy nine-to-five schedule of psychiatrists. It was true. The hours were shorter. But the intensity required to sit with these patients and help them find peace in their lives was overwhelming.

The Power of a Question

Mary had long pigtails that naturally twisted into sausage curls as they fell past her shoulders. She had large, sparkly brown eyes and a quick, shy smile. Mary and her mother spoke only Spanish; they were illegal immigrants. That day Mary had come to the clinic for a regular checkup. When we were done, I gave her a Pocahontas sticker which she hesitantly took from me. "Who's that woman," I asked her in terrible Spanish, using the wrong word for "woman." "Who's that?"

She looked at her mother and then back at me. "Pocahontas," she finally said.

But Mary's legs still hurt her. Just two years before, Mary and her mother lived in an old apartment, and on a regular screening examination Mary's lead level came back at eighty-one, much higher than the upper limit of eleven. Blood levels exceeding sixty brought the risk of diminished intelligence. But since Mary and her mother didn't have a telephone and couldn't read English, it took another month to track them down. Someone actually went to their home to find them. Mary was admitted to the hospital for chelation therapy to remove the lead from her body. X rays showed bony changes in her legs consistent with severe lead poisoning, likely responsible for the leg pain.

The only way to prevent the lead from reaccumulating in her blood was to remove the lead paint from her apartment. But since she and her mother lived there without a legal lease, the landlord refused. Mary and her mother were forced to live in a shelter, and the strict shelter chore requirements caused them to miss their appointments at the clinic. After a long battle the clinic finally forced the landlord to de-lead the apartment, and Mary and her mother returned home just before Christmas, nearly a year later. Now, still without a lease, Mary's mother couldn't provide proof of address to enroll Mary in kindergarten for the upcoming school year.

"You can't enroll her? But she needs to go to kindergarten. This is very important," Dr. Sands told her in heavily accented Spanish. Dr. Sands was one of the few physicians at the clinic who were not Latino, but over the years she had learned Spanish. She was tall and effusive, with carefully sculpted blond hair and glittery jewelry. She was also sensitive to the needs of this community, and her patients loved her.

Mary sat on the examining table, chattering at me in a rapid stream of lilting Spanish that I couldn't understand. "Did you go to the parents' committee?" Dr. Sands asked.

"Yes, but they asked me for a lease, and I didn't have one," her mother said.

"Are you living with someone who has a lease for your apartment?"

"No. We don't have a lease."

Dr. Sands picked up the phone in the examining room and made a quick call to the parents' association. She held her hand over the receiver and asked Mary's mother, "Can you go over right now? Ask for Francesca. I explained your situation, and she's expecting you. She'll get Mary enrolled in kindergarten. But you have to hurry because they close in forty-five minutes."

Mary's mother quickly helped her back into her T-shirt and jumper, and they smiled and nodded at us as they left the office. "Thank you, thank you," the mother said.

Mary and her mother lived in a predominantly Latino suburb of Boston where I did an elective in primary care pediatrics. Since my inpatient pediatrics rotation during the winter, I had become enchanted with pediatrics. It was already June, and my residency choice fast approached—I had to start the application process in only two months—and I struggled to choose between pediatrics and adult medicine. I knew I could be happy in both specialties, so I wasn't worried about the decision. I just didn't know how to make the choice.

Now at the beginning of fourth year I had finished most of the required rotations and had free time to choose clinical electives. Unlike the core rotations I had just completed, electives were only a month long. Some, like the intensive care unit rotations, required long hours and overnight call, but most were less demanding than the core rotations. My classmates and I could select electives in whatever specialties we preferred, and we used these months to explore various careers to help us choose a path in medicine.

While I loved caring for very sick children in the hospital, I wasn't sure I would enjoy seeing healthy kids. If I could enjoy outpatient pediatrics, I reasoned, then I knew I could be happy as a pediatrician. I needed this month to decide whether I could enjoy the outpatient experience.

This Latino suburb had a large illegal immigrant population. Because the community was so predominantly Spanish-speaking, many immigrants did not learn English. One English-speaking mother told me she put her five-year-old son in a bilingual classroom so he could understand the people working in the grocery stores and translate for her. It was not uncommon to have a patient in the office whose parent or grandparent spoke no English, only to learn the family had lived in America for ten or twenty years. There were plenty of Spanish-speaking employers, and even the public schools offered nearly the entire core curriculum in Spanish or "shelter classes," with modified vocabulary for new English speakers.

The clinic was a large, modern facility. It offered clinics in adult medicine, pediatrics, obstetrics and gynecology, and mental health as well as basic emergency services. Nearly all the physicians in the pediatric clinic spoke fluent Spanish, and all the secretaries, nurses, and other support staff were hired from the local community. I had never taken a Spanish course in my life, but I had been trying to learn Spanish from language tapes to speak with Carlos's family. I learned the names of the parts of the body before I started, hoping that would pull me through the month. By the end of the month I was able to perform a very basic history consisting primarily of yes-no questions in Spanish, but I was lost if the patient started to tell me something I didn't expect to hear.

The medical problems were generally not the most prominent issues for the patients I saw at this clinic. A child came in for a sore throat, and in the process of asking a few questions, I learned the family was in danger of losing their electricity or phone. Or there was drug abuse or violence in the family. Or the child was failing out of school. This month impressed on me the importance of asking because I was never sure what I might learn. Often the problem I discovered was more pressing than the issue that brought the family into the office. In my first year of medical school I was introduced to the intimacy of the patient-doctor relationship and the importance of recognizing the human experience of the patients I met. But in second and third years I was caught up in transforming people into patients so that I could best care for their medical needs. This month reminded me of the dual significance of my role to see these children and their families not only as patients but as people.

Gang and drug violence was also prevalent in this community, and not restricted to teenagers. The mother of one of our patients was Latina, with dyed blond hair, bright lipstick, red nails, and a rhinestone belt. Dr. Sands had taken care of this woman's children for years. She had been married to a man involved in gangs and drugs, the father of both her children. She fell in love with an FBI agent involved in arresting her husband. She divorced while her husband was in jail and then married the FBI agent. When her first husband was released from jail several years earlier, she worried that he might come after her or the children. But as far as Dr. Sands knew, everything had gone smoothly. That afternoon she had brought her eleven-year-old son and eight-year-old daughter for school physicals.

Her son and daughter sat on either side of her in the examining room. Her son talked easily with us, telling us about school and his friends. When it came time for the exam, he willingly hopped up on the table and took off his shirt. His sister, however, was extremely shy. She huddled closer to her mother when we came into the room and regarded us watchfully with her deep brown eyes. She wouldn't speak to us. Her mother mentioned that she was having some trouble in school. One year before, they had moved to a more affluent neighborhood with better schools, and her daughter's teacher had sent a note home because the child refused to go to the bathroom by herself.

When it came time to examine the daughter, she ran and cowered in the corner by the door, watching our every movement until her mother helped her go to the examining table. She was reluctant to take off her dress and hugged it closer to her as we threatened to remove it. Finally we were able to coax her to take her dress off, but she startled as I touched her with my stethoscope although she had watched me bring it to her chest.

Dr. Sands reassured her mother. "She's certainly a shy child, but after coming from these schools, I'm not surprised she's afraid of the bathrooms. I would be afraid to go to the bathroom by myself too. She'll probably grow out of this, but if she continues to have trouble, we might need to look into this more carefully."

As I watched the little girl, I was slightly worried. I wondered whether she had been abused. The hypervigilance as she eyed us, the fear as she cowered, and the strong startle reaction to my stethoscope seemed abnormal to me, more similar to the response of people suffering posttraumatic stress disorder. I wondered why she was afraid of the bathroom in particular. Had she been sexually abused? Her exam was normal. She had no unusual marks or bruises and no signs of sexual abuse on her genitalia. After we left the room, I mentioned my concern to Dr. Sands.

"I guess I hadn't really thought about that. She has such a great mom. But it's absolutely possible, and now that you mention it, her behavior could be an indication. I'll have to explore that further in our next visit," Dr. Sands said.

I saw terrible child abuse in this community. Child abuse is not a disease of poverty or immigrants, so there was no reason to expect more child abuse in this community than elsewhere. But family structures were complicated, and most people lived with relatives as they struggled to get on their feet after recently emigrating. Parents often worked long hours to make ends meet, and children had many hours of inadequate supervision. Some of the pediatricians speculated that these extreme social stresses created enough chaos to permit incest, abuse, and neglect.

I nearly bumped into a woman pushing a stroller with two kids trailing behind as she rushed out of the clinic. Dr. Sands pulled me aside to tell me their story. "Did you see that woman and her three kids? They always come hours past their appointment here and late for their mental health appointment upstairs. The middle child has attention deficit disorder, and several years ago he fell out a third-story window, fractured his skull, and spent a month in the intensive care unit. She started coming to our clinic about a year after that happened.

"But then, a couple of months ago she came for her appointment with the older and the younger child, but the middle child was missing. So I asked her where her middle child was, and she said, 'Oh, I couldn't bring him today. He can't get up.' 'What do you mean, he can't get up?' I asked her. And she told me that the night before, he had been climbing a tree and had fallen, essentially impaling himself on an iron fence. So I of course immediately called the ambulance and DSS [Department of Social Services]. It turned out he had a fractured pelvis and a huge collection of blood in his scrotum. They drained the scrotum, but they didn't stitch it up and left it to heal on its own. So it was important that the kid take a bath every day to keep the area clean and from getting infected. Today I asked the kid if he was taking a bath, and he said no. So I asked his mother, and she said she couldn't make him take a bath. I reminded her why it was so important, and she said, 'Well, what do you want me to do, beat him?' "

Dr. Sands told me about her most painful experience. A mother brought her fourteen-year-old daughter to the clinic because of stomachaches. She'd been getting fat too, the mother complained. Dr. Sands looked at this girl and realized immediately that she was pregnant. She had a hard time getting the mother to leave the examining room but finally forced her to leave for the examination. The girl undressed and lay on the table. "You know, you're pregnant, and you're pretty far along," she told her. "Do you have a boyfriend?"

"No," the girl told her. "My mother doesn't let me have boyfriends."

"Well, you must have had a boyfriend once," Dr. Sands said.

"It's my daddy."

"I don't mean your father. I mean, a boyfriend. You must have had a boyfriend once," Dr. Sands repeated.

And the girl told her again, "It's my daddy."

"She had to tell me three times before I finally realized what she was saying. I was so horrified I just had to leave the room to give myself some time to think of what to say. As I walked out of the room, the mother was sitting outside, looking at me like, 'Can I come back in now?' I wanted to vomit," Dr. Sands said. "It turned out that the mother worked until five, but the father came home at three. He molested the daughter every afternoon after school."

Brian was a thirteen-year-old boy, in the midst of the gangly awkwardness of his growth spurt. He had strawberry blond hair and angular features, and he wore a black T-shirt with baggy jeans set low on his hips. He was pleasant, but he shifted rapidly between impish exuberance, playing with tongue depressors and the ophthalmoscope, to a dark scowl, staring at the corner of the room. He had been diagnosed with attention deficit disorder many years earlier. That afternoon he came with his parents to the clinic because his family was in crisis, and he was in the middle of court hearings to place him in foster care.

Brian's parents were divorced, and his mother's boyfriend hated Brian. They fought bitterly, and he no longer allowed Brian in their home. For the past eighteen months he had lived with his father and his sister. A severe asthmatic, Brian had begun smoking this year. He acquired the habit from his father and sister, who supplied him with cigarettes. Smoking can be a significant trigger for asthma attacks. Brian had been admitted to the hospital one week previously for a severe asthma attack, and in the emergency room doctors first learned that he had been truant from school for the last year. Apparently Brian had decided not to go to school, and his father allowed him to stay home and play video games all day. For some unknown reason, the school had not followed up on his absences. Brian had been seen in clinic by Dr. Sands once in December, but in an uncharacteristic slip she had not asked him about school.

"I can't believe how stupid of me. I always ask, and of all the times to make a mistake. I feel terrible. Maybe I could have averted all this," she said.

Brian was placed in foster care that day, and we saw him back in clinic on Monday to find out how his weekend in his foster home had gone. When we came into the exam room, he sat on the exam table, fidgeting and turning the ophthalmoscope light on and off.

"It was pretty much okay," he told us. "There's lots of other kids there, but Laura made us do all these chores. We all played basketball in the afternoon. It was fun."

Dr. Sands knew Laura. "She's one of the better foster moms. She's strict, but she maintains control of the kids and keeps them safe. I'm glad he's there." Although the duration of his stay in foster care was not specified, he would stay indefinitely with Laura. DSS arranged for him to start eighth grade in the fall.

All these children needed their caregivers to ask. They needed us to discover what was happening to them in their homes. For some, we didn't have the opportunity to ask until it was too late. For others, we inadvertently missed our opportunity. But sometimes our questions and encouragement could make a difference.

Sarah came to see Dr. Sands one afternoon to get a prescription for acne medication. She was very petite, much shorter than I am, and dressed in a suit. "Wow! You look great," Dr. Sands said.

Sarah beamed. "I'm still too skinny." She proudly showed us her heart-shaped diamond ring. "I'm engaged now," she told my preceptor. "He works in a garage, and we're getting married next year."

"Congratulations!" Dr. Sands said. "And I heard that you were coming to work here as a secretary. When do you start?"

"Today was my first day of orientation. I'll be working here, in the pediatrics department."

"And your sister?"

"She's doing great. She finished travel agent school and got a job working for a law firm downtown. She commutes to Boston every day."

Sarah needed a simple prescription, so we didn't keep her in the office very long.

"I met Sarah and her sister three years ago," Dr. Sands said. "They had dropped out of high school, and they were living with their parents in an apartment that someone in their family owned. None of them had a job, and they were literally starving. But in the last two years they've picked themselves up and turned their lives around. You'd never know this was the same person to look at her."

I loved my month in outpatient pediatrics. The medical issues were generally less complicated than the problems I encountered in the hospital. But outpatient care had its own challenges. In the hospital, support for a sick kid was immediately available. But in the outpatient setting the repercussions of a bad decision were much scarier.

I especially loved contending with the social problems. These patients needed their caregivers as advocates to make sure their basic needs were met. The children who weren't old enough or sophisticated enough to protect themselves needed a voice to speak for them. The parents who wanted the best for their children but had no concept of American standards needed someone to educate them about reading, vaccinations, diet, and regular physical exams. I relied on all my resources to care for these children: my skills as a medical professional, my powers as an advocate, my caring as a human being.

This month convinced me to choose a career in pediatrics. In some ways I think I have chosen stories over medicine. I might sacrifice some medical complexity for greater emphasis on social and developmental history. Interestingly, I have chosen the same residency as Perri Klass and Claire McCarthy, two other Harvard women who have written about their experiences in medical school, and Neal Baer, the Harvard graduate who writes for the NBC-TV series *ER*. I can't help but think that our strong commitment to stories has led all four of us to choose a career that honors the personal and family history.

Hazard

It was our first day in the hospital. Carlos and I wandered around the nurses' station, trying to figure out where to write orders for our patients. A young man in a pin-striped shirt and red necktie kept looking over at us as he doled out medications into the individual trays for each patient. Finally he came over and introduced himself. When we told him we were medical students, he said, "I thought so. You're the students from Harvard, right? Everyone's been saying that you were coming. You know, I just have to ask because we've all been wondering. How the heck did you end up in Hazard?"

Hazard is in the Appalachian region of eastern Kentucky. With a population of nearly seven thousand, it is the third-largest town in the area. The next town over is Typo. Hazard is built right in the mountains; there is no discernible valley, and flatland is a valuable commodity. One of our residents said about Hazard, "It's as if someone just got tired of walking, plunked himself down, and said, 'Here. I ain't walkin' no further. Let's build here.' "

Carlos and I ended up in Hazard by accident. I had always had a fantasy about living in a rural community. As a child I loved animals and, fascinated by James Herriot, believed for many years that I would become a big-animal country vet. By the time I started medical school I had long given up on my veterinary aspirations. But the idea of living in the country among farms and animals still lingered. I thought back to Dr. Abraham Verghese's book *My Own Country* detailing his experiences caring for AIDS patients in Appalachia. I had read the book two years previously, but the characters and stories made a lasting impression, and I thought I might also enjoy working in that area.

As I remembered, Dr. Verghese had worked in Kentucky. I found a phone number for the National Center for Rural Health, and through a telephone tree I couldn't even begin to trace back, I wound up with the number for the family practice clinic in Hazard, Kentucky. I bought Carlos a copy of Verghese's book so that he could get excited too. Carlos turned over the book and read the first sentence on the back of the jacket: "Nestled in the Smoky Mountains of eastern Tennessee, the town of Johnson City had always seemed exempt from the anxieties of modern life . . ."

"Hey," Carlos said, "this isn't Kentucky. It's Tennessee. You've got the wrong state!" We were not off to a good start.

We made the two-day trip from Boston to Kentucky at the end of July. At the entrance to Hazard we passed a tall tower bearing a large sign that proclaimed HAZARD: QUEEN CITY OF THE MOUN-TAINS! Main Street ran through the middle of downtown, which consisted of two parallel roads approximately a quarter of a mile long. Rows of worn 1920s brick buildings lined the streets, some adorned with faded plastic signs that looked to have been added in the fifties. Hazard was once a bustling coal town, but now most of the storefronts were empty. Handwritten signs dated from January were still posted in empty display windows, advising customers that the store was moving to a new location. Most of the businesses had moved away to the malls on the outskirts of town, built on the scars of old strip coal mines.

Neither Carlos nor I had ever lived in such a small town before, and it was quite a contrast with Boston. In all Perry County there was not a single bookstore, and Hazard had only one nonchain restaurant. The city closed down at four-thirty every day, and nothing was open on the weekends. One Saturday morning the popular radio station held a four-hour live broadcast of the back-to-school sale at Wal-Mart.

I had just finished working in the Latino community clinic, and I was interested to compare the experiences of urban and rural poverty. I expected the medicine in Kentucky to be as different as the scenery. This world, so utterly unlike anything I had ever experienced, must have its own unique illnesses. This community, living in such a different environment, must have its own set of social issues. Carlos and I weren't sure how this community would respond to us, and we feared they would regard us as two voyeurs from Harvard here to see their poverty. We feared they would be reluctant to enter even a medical relationship with us.

I was surprised to find that the health effects of poverty were similar in Boston and Hazard. Poor patients in both communities had difficulty accessing primary care and, as a result, tended to suffer severely with common illnesses, such as diabetes. In both communities many people went underinsured or uninsured. In Boston it was more common for extremely poor families to be homeless, while in Kentucky even the poorest had homes, but they might not have running water or electricity. The health summary questionnaire given to all patients admitted to the Hazard hospital, asked, "Do you have an indoor bathroom?" Patients could check "yes" or "no."

The hospital in Hazard was remarkably large given the size of the community. But it served as the primary referral center for five counties. A new brick and glass building built on an old strip mine, the hospital looked very much like the Harvard hospitals I was accustomed to, although on a smaller scale. It had an eight-bed cardiac intensive care unit and an equally large medical intensive care unit. It offered care in all the major specialties except neurology and cardiothoracic surgery. If a patient was too ill to be cared for in Hazard, the hospital offered helicopter service that could transport him or her to the University of Kentucky hospital in Lexington in fifteen minutes.

The biggest difference between Boston and Kentucky was not the degree of poverty or the medical resources available, but coal. Coal was the predominant industry in the region, and the mines operated twenty-four hours a day, divided into three eight-hour shifts. Nearly all the men were connected in some way to the coal industry, and many of the men developed black lung, in which years of inhaled silicon stiffened their lungs. The mines demanded hard work with heavy machinery, and many of the men had been in mining accidents. Mining was hard on the body, and the men looked old beyond their years by the time they retired in their early fifties.

Coal created an unusual economic and social system. Winter conditions and long commutes made getting to school difficult, and mining did not require a sophisticated education. Conditions had improved over the years, but especially among the older generations, illiteracy was prevalent. A few men with virtually no education became tremendously wealthy by working their way up the mining hierarchy, particularly during the oil embargo of the 1970s. Nancy, one of the nurses, told me that her neighbor had become a multimillionaire from mining, but he was illiterate. He used to bring his mail to her father, who had a third-grade education, to read.

Hazard and eastern Kentucky also had the highest smoking rates in the country. Nearly 50 percent of adults smoked. While silicon exposure from the mines has never been proved to cause lung cancer, the miners' lungs were already severely compromised when they developed lung cancer from smoking. In Hazard, 40 percent of all cancers diagnosed were lung cancer, compared with the national average of 14 percent. A resident who was from the area said, "Around here these guys just go to cancer." Getting lung cancer was not an event, just a matter of course. During our month our medical team diagnosed at least one and more often two new cases of lung cancer each week.

One of my patients had recently lost her husband to lung cancer. In her early sixties, she was thin, and her white hair, yellowed with age, was pulled into a tight ponytail. Her maroon-tinted glasses magnified her blue eyes and made them seem too large for her face. She had married her husband when she was just fourteen, and he was twenty-one. "When he went to ask my daddy to marry me, my daddy told him I wasn't nothin' but a little girl. 'That's okay,' he said to my daddy. 'I'm gonna raise her up my own way.'" She said they'd had a happy life together, and she sorely missed him now that he was gone. "I remember when he used to come home from work. He was always lickin' his lips. He would come home from the mines with his whole face black with coal dust except for a little circle around his lips where he'd licked it all clean."

I was thrilled with our experience in Hazard. The hospital was a friendly place, and I enjoyed the other doctors and residents. Specialists often stopped Carlos and me in the hall to give us an update on patients we shared, and they sat with us at lunch, something unheard of in the hierarchy of the Boston hospitals. Hazard was not an academic center, and the residents did not read the current literature as much as we were used to in Boston. But medical care was excellent, and I had confidence in all the residents and staff physicians. My patients were challenging, because of both their complicated medical issues and difficult social situations. I loved experiencing another culture and lifestyle.

Carlos and I had a very special month together. We worked together all day and spent our weekends and evenings without the distraction of friends, a TV, or a phone. I was happy to spend the concentrated time with him, and it reaffirmed how excited I was to be marrying Carlos. We went for a long walk every evening and hiked every weekend. I realized how much I had missed nature after living in New Haven and Boston for the previous seven years. I think Carlos missed the bookstores, restaurants, and bustle of Boston more than I did. Although I would eventually miss the city, I felt I could be happy living in a small community for several years. Our experience in Hazard made me want to return to work in this community after we finished residency.

At the end of our month in Kentucky Carlos and I took the second part of the Boards, the national medical licensing examination. The Board examination was the last major hurdle to finishing our medical education. There were no further exams required to earn our degrees. Like part one of the Boards, the second series in the Board examination spanned two days. It covered all the basic medical specialties and questioned us on diagnoses and medical management of diseases. Having passed the first part of the exam, I was more relaxed about the second part. My study sessions were less intense and much more abbreviated. During the last week of our rotation in Hazard, Carlos and I drove two and a half hours to Lexington, Kentucky, the nearest testing site. We felt adventurous taking our Boards in an unfamiliar place. After the Boards the last obstacle standing between us and graduation was the match, the centralized residency selection program that matched medical students to residency positions in hospitals across the country. The residency application process began in early fall and stretched through the year.

Carlos and I returned to Boston in early September anticipating a relaxed year ahead. With most of my hard rotations out of the way, I had filled my schedule with lighter clinical electives. After returning from Kentucky, Carlos and I moved in together. We rented a sunny one-bedroom apartment in our old neighborhood in Brookline. Much as I loved my old basement abode, it was a relief to live in a bright apartment looking out on green summer trees.

Our friends also had lighter schedules. All of a sudden we had our evenings and weekends off, and we began to socialize again. One Saturday night we gathered at Roy's apartment and sat talking until one o'clock in the morning. I realized it had been well over a year since we had stayed out that late. For the first time in months no one was postcall and exhausted, and no one had to get up early to be in the hospital the following morning.

ER Reprise

As we approached the deadlines for residency applications that October, rumors of who had chosen which specialties dominated our conversations. At a bridal shower for one of our classmates I realized how obsessed we had become. Her fiancé was an actor, and his friends talked animatedly, eating and laughing, while we medical guests sat in a corner weighing the various hospital programs and trading advice.

At first it was strange to identify myself and my classmates with particular specialties. I was so accustomed to thinking of us as undifferentiated and unskilled, with every option still a possibility. Now that I had chosen my future and begun the application process, however, internship and residency suddenly became concrete. I would be a doctor soon.

There were a few surprises in the choices people made. Alyssa, who we all had been certain would choose a surgical specialty, chose pathology instead. One afternoon midway through third year our Patient-Doctor session was devoted to a game. We each turned in a secret ballot listing our classmates and what specialty we thought each would choose. All of us received a mixture of responses. Half my classmates pegged me as a general internist; one person thought I would be a geriatrician. But for Alyssa, the response was unanimous. Her list read: "Surgeon. Surgeon. Cardiothoracic surgeon..."

I remembered her telling us once that she envisioned herself working with critically ill patients alongside her husband. She imagined them grabbing ten-minute dinners at 2:00 A.M. in the cafeteria during the occasional free moments. But over the summer she did a clinical elective in forensic pathology, and she loved poring over the cadavers, searching for any clues to the mechanism of death.

"And I want to be a mom and have time to spend with my kids. That's just not possible in surgery," she told me as we walked home after Patient-Doctor late in September. Her husband had chosen to specialize in surgery, and juggling two surgical careers with long and often unpredictable hours would compound the issues of raising a family. You couldn't duck out of a ten-hour surgery to pick up your child sick at school. Alyssa had taken several years off before medical school, and she felt her biological clock ticking. While some residencies were known to be flexible about parental leave and offered part-time positions, the surgical programs were not among them.

As I thought back over the years I had spent with Alyssa in Patient-Doctor, her choice seemed less incongruous. At times she had felt frustrated by the human responsibilities that crowded her time and made it more difficult to focus on the disease mechanisms and the therapies. After a practical exam during second year in which we had two minutes to examine each of thirty "patients," she had commented, "It was great because I could just go in and focus on the problem without all the small talk."

At other times she was paralyzed by the grief she felt for her patients. I remember her crying in Patient-Doctor as she told us about a dying patient in the hospital. The elderly patient was in pain, and the physicians had given her so much pain medication that she was too sedated to be emotionally present with her husband as they shared their last moments together. "He never got to say good-bye. He never got to say 'I love you' that one last time," Alyssa said.

"I felt like I would need to carry my beeper twenty-four hours a day and be available to my patients at any moment to be the kind of surgeon I wanted to be," Alyssa told me after she made her choice. "When it came down to it, I didn't want to give up that much of my life."

Although I appreciated her reluctance to sacrifice family for career, I was disappointed by her choice. Only two women in our class had chosen general surgery, in large part, I think, because of the time-intensive schedule. Alyssa was one of the few people, male or female, who I knew could keep up the grueling pace. She would have been a wonderful surgeon. She was a dedicated student and had amassed an impressive knowledge of medicine. While she had her moments when she could be difficult to deal with, she was often sensitive and insightful about her patients and colleagues. She had the courage to stand up for her beliefs and the strength to interpose her spirit into a hierarchy that could be tremendously intimidating. Alyssa would have been a humanizing influence in what I had experienced as a barren environment.

There were other surprises, which in part reflected the growing trend toward primary care. In a school well known for producing subspecialists, only nine people chose general surgical residencies, and an unprecedented ten students chose family medicine. The number of students applying in pediatrics swelled to nearly thirty applicants. Many others chose primary care–focused general adult medicine residencies rather than subspecialty-oriented programs.

Most of my friends chose what I had expected. Roy selected adult internal medicine in a primary-care-oriented program. Like Carlos and me, my roommate, Kate, opted for pediatrics. Nearly half our class decided to take an extra year to finish medical school. Andrea decided to postpone graduation for an extra year to write a book about her experiences working with women who sell sex in the legalized brothels of Nevada. Masha also decided to postpone residency for a year to study in Israel.

Renu, who had taken an extended medical leave after experiencing severe depression and despair the previous winter, returned to medical school that fall. She still struggled with her decision to pursue a career in medicine but was determined to finish medical school. I bumped into her wearing her Harvard Medical School white coat in the hospital. "It's weird to be back, but I'm doing okay. I'm starting off slow with a few easy rotations to work my way back into things," she told me. I gave her a quick hug before we both rushed off to our afternoon clinics.

Fourth year was a relief after the intensity of third year. I continued to work hard during my days in the hospital, but I had no overnight call and virtually no weekend responsibilities. Carlos was less busy as well, and we enjoyed the extra time to spend together. We started planning for our wedding in May, just two and a half weeks before graduation.

The residency applications were not tedious, unlike the college and medical school applications with multiple sections and dozens of essay questions. I wrote only one personal essay that was applicable to all the schools. Interview season for most specialties did not begin until November and peaked in December and January. We were given a month of vacation to complete our interviews. Match day, when we would learn our residency placements, was not until March 18. In this process, unlike the college and medical school application processes, we would not receive multiple acceptances that we were free to choose among. Instead students submitted their ranked lists of programs to a centralized agency, while the hospitals submitted their ranked lists of students. A computer then matched students to the highest-ranked programs that had also ranked them. On match day, at exactly noon, we would each receive a white envelope with a single piece of paper listing the hospital and program where we would spend the next several years. There was no opportunity to change our placements once they had been given.

Carlos and I participated in the couples match. This meant that the computer linked our applications together. It made submitting our list a little more complicated, but otherwise the system was the same for us as for our classmates. However, because the computer considered couples together, couples were limited by the weaker partner. There was plenty of advice for couples matching in different residency programs, but Carlos and I, matching together in pediatrics, were in a more unusual situation.

"You definitely shouldn't tell people you're couples matching," one adviser told us. "People are going to wonder why you want to enmesh your private and professional lives. Especially in smaller programs they're not going to want a couple."

Despite this adviser's recommendation, most couples disclosed their situation during interviews that went well. If a program liked you enough, the director might call your partner's program to encourage that program to accept him as well. Occasionally a student in a couple received a better placement than he would have achieved on his own.

When we went to our first interviews in December, Carlos and I had no idea what we should tell our interviewers. We finally settled on a don't ask, don't tell policy. Yet while interviewers generally adhered to the "don't ask" portion of the policy for Carlos, nearly every single interviewer asked me. In an informal poll of my friends I found that women were far more likely than men to have been asked about their relationship status and family plans.

At one hospital I had a late-afternoon interview. There were nearly fifty medical students interviewing that day, and the administrators had organized the interviews in a wing of examining rooms. Each interview lasted exactly twenty minutes. Each of us entered one of the examining rooms, where an interviewer with our file in hand waited to question us. After fifteen minutes, an administrator walked down the hall, rapping on doors to remind the interviewers that the session was nearly over. A second knock on the door signaled the end of the interview. We simultaneously walked out the doors and swapped examining rooms for a second set of interviews.

My interviewer had a pile of four manila folders on the small desk in front of her. She was pregnant, with honey blond hair piled into a loose bun. I was the last interview of the afternoon. "I'm sorry, but I didn't have a chance to read your application. I wasn't expecting to do this this afternoon," she said as I came in.

After telling me a little bit about her program, she noticed my engagement ring. "Oh, so you're getting married," she said. "What does your fiancé do?"

I was a little uncomfortable with the conversation, but I told her that Carlos was also a medical student. She pressed further, and I finally told her that Carlos and I were both matching in pediatrics and preferred to be in the same program rather than in different hospitals in the same city.

"Oh, really," she said. "Have you thought this through carefully? I think it would be too claustrophobic to work in the same program. You'll be so engrossed in each other's lives."

Now I felt distinctly uncomfortable. She, not I, had brought up my marital situation, and she was questioning me in detail. She was interviewing me for a job. Wasn't this illegal?

She was not the first interviewer, or the last, to ask me about Carlos. I finally came up with a pat explanation about choosing the same profession, reassuring my interviewers that there would be nothing to worry about if they took us together in their program.

At one of my last interviews I was interviewed by a pediatric cardiologist. "Let's go next door to my husband's office," she said when I arrived. Not only were they both pediatricians, but they practiced the same subspecialty. She was warm and immediately put me at ease. Now here was someone I *wanted* to tell about Carlos. So I launched into my usual explanation, detailing our experiences working together during our subinternship and in Hazard and explaining how happy we had been.

She looked at me across the desk, a little confused. "Of course. Why wouldn't you want to be in the same program?"

Fortunately, Carlos and I had chosen to limit the number of

interviews we would do. Traveling was expensive, and the days filled with multiple interviews and information sessions were exhausting. I was already complaining after only four, and most of the people I met were on a thirteenth or fourteenth interview. Instead Carlos and I spent the time and money we would have used for extra interviews to take a trip to Argentina. His immediate family had lived in the United States for many years, but his grandparents and other relatives still lived in Buenos Aires. It was thirteen years since he had been back, and most of his relatives would not be able to come to our wedding.

After six weeks of interviewing and traveling, Carlos and I finally returned to Boston in February, ready to start our final rotations before finishing medical school at the end of April. We were finally in the home stretch.

In February I began a monthlong rotation in pediatric emergency medicine at the hospital where Carlos and I hoped to do our residency. I thought it ironic that my final clinical encounter of medical school was in the emergency department, as was my first vicarious experience of medicine, through the *ER* TV series. Four years earlier I had identified with the character of Carter, then a medical student, as he blundered through his first patient experiences. My first clinical rotation during third year was also in the ED. I had waited awkwardly for guidance from residents and staff until I could accrue enough skills to become a functioning member of this new community. Now, nearing the completion of my last year of medical school, I was well versed in the rhythms of the hospital and well acquainted with patient care. Finally I was on the verge of becoming a real doctor. I belonged in this world, and I had worked hard to earn my legitimacy.

The pediatric emergency rotation was a calculated choice. Carlos and Kate both had raved about this rotation. Medical students had a significant role in patient care, essentially functioning as interns, and there were plenty of opportunities to learn procedures.

Learning procedures was my goal. I had spent several months working with children and learning how to examine them, but I had never once tried to draw blood. I viewed their delicate veins with trepidation. I was terrified of their pain and their parents' anguish. But residency was bearing down on me. What would I do when I was alone in the hospital in the middle of the night and had never learned to place an IV on a child? I needed to learn procedures, and I needed to learn them fast.

Working in the same hospital where I hoped to pursue a residency added an element of stress. The hospital would formulate and submit its ranked student lists that month, and although logically I knew this rotation was too late to affect the choices the selection committee would make, I still worried.

The emergency department was in an older section of the hospital. It was large, with two hallways and more than thirty patient beds. Despite its size, the department was often overflowing with patients, resulting in long waits. The hallways and rooms had faded over the years to a uniform nut brown, and the dim yellow lighting made the small rooms seem even smaller. Unlike the other areas of the hospital, which were brightly colored, cheerful, and well stocked with toys, the ED gave little indication of the children it cared for. Only the occasional stray miniature yellow johnny gave it away.

My first week in the emergency department was great. I followed my own patients as they came into the department, formulating a differential diagnosis for their problems, organizing their diagnostic tests, performing their blood draws, and taking primary responsibility for tracking their progress through the ED to discharge home or admission to the hospital. I worked directly under the supervising staff physicians without a resident following our patients with us.

This was almost exactly what being an intern would be like. As I took care of my patients during the first week of the rotation, my confidence surged. I was able to focus my clinical exams and develop reasonable treatment plans. I was able to keep track of the details for each patient. While not yet successful at blood draws, I was overcoming my fears of learning. After four years I finally felt I had acquired enough skills to be an intern. I could do this.

But the second week was a disaster. On Monday I arrived at 11:00 A.M. ready for a twelve-hour day. My first patient was a two-and-a-half-year-old Haitian boy who had come in with his mother. His mother was frustrated. Her son had had episodic diarrhea for the last month. Each time she took him to her pediatrician, he was diagnosed with viral gastroenteritis—stomach flu—and sent home without medication. Over the weekend he again had a few episodes of diarrhea. But he became fussier on Sunday and developed a fever late in the evening. His mother brought him to the ED on Monday because she was fed up with her pediatrician.

The little boy sat in his mother's lap, playing quietly as she talked. He didn't look seriously ill or even particularly dehydrated. The only notable finding on exam was an ear infection. Dr. Jameson, the senior physician, reviewed the case and agreed with my assessment. I had not worked with her before. She was young and energetic, with an exuberant voice and long black hair.

When I first interviewed the boy's mother, she told me that he had developed a mild, nonitchy rash in allergic response to amoxicillin, usually the preferred antibiotic for ear infections. I briefly reminded Dr. Jameson of the allergy as we discussed antibiotic choices. "He has an allergy to amox. D'you want to use Bactrim?" I asked.

She thought for a few seconds before answering, "No, let's use ceftriaxone IV, since he's already got an IV and one dose is the full course of treatment."

I knew that a small minority of patients with a severe, life-threatening type of reaction to amoxicillin were likely to be allergic to the ceftriaxone as well, but patients with the skin rash allergy his mother described did not usually fall into this category. I had seen other physicians use ceftriaxone despite this contraindication. Plus, I had just reminded Dr. Jameson about the allergy.

So, if she still wanted to use ceftriaxone, I was comfortable relying on her judgment. I wrote the order for the ceftriaxone, and a resident cosigned the order so the nurse could administer the antibiotic.

Just a few minutes after the nurse told me the patient had received the ceftriaxone, Dr. Jameson came up to me in the hallway outside his door. "You told me he had a Bactrim allergy," she said.

My stomach dropped, and sick sourness welled up in my throat. "No, I think I told you he had an amox allergy. I was *suggesting* Bactrim," I said.

"No." She raised her voice slightly. Her furrowed brow and intense eyes burrowed into my soul. "You told me he had a Bactrim allergy."

"Well, I'm sorry if you heard that. But I really thought I told you he had an amox allergy," I replied. My heart raced, and my hands were cold and clammy.

"Well, you just can't do that. You can't give patients with an amox allergy ceftriaxone," she said angrily.

"I didn't ask because I've seen people give ceftriaxone under these circumstances, and I thought you'd heard me," I said shakily.

"But this is serious. This kid could have a potentially fatal complication. You should have asked if you had any questions." She raced off to the boy's room, and I rushed to follow her. In a brusque, anxious manner she questioned the mother again about the boy's allergy. The mother explained the rash all over again.

When we walked out of the room, Dr. Jameson said gruffly, "Well, that kid will probably be okay, but we'll have to watch him more closely."

I felt horrible. I couldn't sit still, and I anxiously paced the hallways of the emergency department as I waited to see if he would develop an allergic reaction. I still had more than half of my shift left, and I didn't know how I would possibly make it through. This was my worst fear realized. I had made not only a simple error of oversight but an error of judgment.

In the end the child was just fine. He did not have an allergic reaction to the medication. When I checked him before he left, he was sucking on a Popsicle and playing with a sand table in the playroom. The child was all right, and Dr. Jameson was all right. But I was a mess.

Too numb for tears, I called Carlos. I had planned to quit my shift early and leave after the patient went home, but I was terrified of the twenty-minute walk home alone. Instead I asked Carlos to meet me at the hospital and walk me home. Carlos met me in the ED, and as we walked to my locker, I explained the situation.

"That's it?" he said. "And the kid's okay?"

I nodded.

"I would've definitely given the ceftriaxone too, under those circumstances," he said. "I've seen ceftriaxone given to kids with mild amox allergies tons of times, and I've done it myself. God, when you called, I thought something terrible had happened."

This was one of the moments I was most grateful to share medicine with Carlos. No one else could calm me as he could, and I knew he understood both the medical and the emotional elements of my situation. These were not empty words of confidence from someone who loved me and therefore implicitly trusted my medical skills.

Instead of walking home, we went to the Au Bon Pain in the hospital lobby and Carlos sat with me while I picked at a bagel. He reassured me again and again that he would have done the same thing in my situation. Finally I regained some composure and went back to the ED to finish my shift. Signing my name in the "doctor" space and taking responsibility for another patient that evening were among the hardest things I had ever had to do.

I knew the antibiotic choice itself wasn't a huge error. But for the first time my potential to make a serious error was real to me. I had come close to making other, worse mistakes. We all had. But this time the mistake had happened. I didn't want this dizzying responsibility.

I had never once doubted my choice to become a doctor in the past three years, but now, just as match day and residency were at hand, I desperately wanted to be anything but a doctor. All of a sudden I started looking at people in other professions with envy. Why couldn't I have chosen a simple nine-to-five job with weekends and holidays off? I had already signed a binding contract for my internship year when I submitted my match list. There was no turning back now. Carlos rushed to keep up with my rapidly fluctuating moods and to remedy my paralyzing crisis of self-confidence.

Time seemed to slow to a crawl as match day approached. Our administrators told us we should line up at the registrar's office promptly at noon on match day to receive our envelopes. They would distribute a class list with the residency assignments after each of us received our letters. Afterward the administration would host a luncheon for us and our families.

"That's like public humiliation," my mother said when she heard how match day was organized.

In our small community of only 150 people it was difficult to keep secrets. Even if I didn't personally talk to each of my classmates, among all of my friends and friends of friends, I heard about nearly everyone. As a result, we knew one another's choices. When we opened our letters together, it would not be difficult to tell who had done well and who had not.

On Black Monday, two days before match day, students who had not been selected by any program to which they had applied would be called by their society administrators. Tuesday was scramble day. On scramble day, society administrators would call program directors at hospitals that still had empty spaces and ask them to take on these students. Virtually everyone was able to get a residency spot by scrambling, although they could be placed anywhere in the country. If we did not receive a call on Monday, then we were certain to have placed at one of our selected hospitals.

On match day morning I struggled to concentrate on my morning lectures, but the blood pulsing through my ears threatened to drown out the speakers. I met Kate, and we walked over to the medical quadrangle a little later than I had intended. Carlos had the day off from his rotation, and we had arranged to meet a few minutes before noon. Certain that he would have arrived already, I commented to Kate, "I just hope he opens his envelope first. Then I don't have the stress of waiting in line myself."

"But you don't think he'd open it without you, do you?" Kate said.

"I really hope he does."

By the time I arrived, jubilant classmates who had already opened their letters thronged the hallway. Pink and white helium balloons festooned the stairway, and I had to fight the crowd to make my way to the registrar's office.

I found Carlos standing on the landing of the stairway clutching his envelope. "Where were you? I've been waiting," he greeted me.

"Didn't you open your envelope?"

"No, I've been waiting. Go get yours," Carlos said.

"No, you open yours first," I told him.

"Are you sure?" he asked. I nodded, and he tore his envelope open. He looked, smiled, and showed me the paper. He had gotten his first choice: it meant that most likely I had as well. I left Carlos to find our other friends and walked calmly into the registrar's office to retrieve my envelope. My letter confirmed that we both had gotten our first-choice placements.

I quickly found Kate and learned that she too had placed at the same hospital. We would be residents together. Roy was also happy with his match. He would stay in Boston as well.

I'm not sure what I expected from that moment, but it was not a euphoric moment for me as it seemed to be for some of my other classmates. I just had a tremendous sense of relief from the tension that had been building over the past months.

A friend who had taken a year off but came to celebrate match day with us told Carlos and me afterward, "I watched you and Carlos open your letter, and you both just kind of smiled and shook your heads. I was wondering if maybe you hadn't gotten your first choice after all. As I looked around, some people were so ecstatic about what they got. I can't imagine finding anything in that envelope that would make me that happy."

Both my parents and Carlos's parents were anxiously awaiting news of the results. I phoned my mother at work, and fighting back tears that surprised me, I told her about us and the rest of our friends. I think she was more excited than we were. I called Carlos's parents at home as well, and they were also pleased we had gotten what we wanted, although disappointed that we wouldn't be coming to California, closer to them.

When I returned to the celebration, people had shifted from the halls into the large conference room for a buffet lunch. As I looked around, everyone seemed thrilled. But as I looked more carefully, I noticed that a few people were missing. We were not obligated to open our envelopes in the hallway in front of everyone, and some people had secreted theirs. I grew concerned as they took longer and longer to return. Once I received the class list with everyone's assignment, it was no mystery why those few hadn't come back to join us for the celebration. While they had received placements, they definitely weren't in the residency programs they had been hoping for.

As one student opened his envelope, three of his friends stood by, snapping pictures to preserve the moment. He seemed disappointed, but he was a notorious joker. His friends called out, "Come on, stop fooling around." They quickly realized that this time he wasn't teasing.

One of my friends arrived late, and as he pushed his way up the stairs to the registrar's office, another classmate who had already seen the class list approached him, "Hey, I know where you're going. Do you want me to tell you?"

This felt like far too public a forum in which to confront our futures. For those of us who were happy, it was acceptable to share the moment. But I felt the pain of my bright, talented classmates who didn't get the placements they deserved. Why should they be subjected to our scrutiny in their moment of defeat? Why couldn't we have privacy to come to terms with our expectations and then return to celebrate the culmination of our four years with our classmates?

Match day was the final hurdle before graduation. Now we could finally envision ourselves as the residents we would be come July 1. I savored the release from tension that match day brought. But the public embarrassment of my less successful classmates marred what could have been a marvelous celebration.

Now that residency was identified and concrete, the fears stirred up by my emergency department experience came to the forefront in a fresh wave of pain. While I knew my classmates were anxious about internship, I was panicked. Carlos reassured me, and I tried to push my fears aside and enjoy my last months of fourth year.

Graduation

After the match the year quietly ground to a halt. I had finished all my clinical hospital-based courses and had only two months of lecture courses left. I had seen the last patient of my medical school career in February. Carlos also had only light course work left. We read more, cooked more, and even went to a few concerts, something we hadn't done since second year.

Now that Carlos and I knew where we would be come July, we could relax and focus on planning our life in Boston. We loved our Boston apartment, and now that we knew we didn't need to move, we got rid of some of our shabby student furniture and transformed our apartment into a permanent home.

We hardly noticed our last day of medical school. I felt as if I had slept through the last month. My lecture course ran only three or four hours on most days, and I spent many of those hours crocheting yarmulkes for the wedding party and half listening to

boring lectures. Both Carlos's and my parents called to congratu-
late us and asked how we had celebrated. But we felt singularly
untriumphant, and our only nod to the occasion was sharing a
beer before dinner.

Carlos and I had reserved May, the last month before graduation, for vacation, as had most of our classmates. Our wedding
was planned for May 17 in Rochester, New York, my hometown,
and we decided to honeymoon in Ireland afterward.

We left for Rochester at the beginning of May to help my
mom with all the last-minute details for the wedding. Although
we had planned a small, simple event, the errands were endless
nonetheless. Initially I had difficulty focusing on our wedding.
While I had regained some composure since my ED rotation, I
still felt a gnawing panic about the impending residency that
overshadowed my excitement about our upcoming wedding. This
background anxiety occasionally spilled over into an acute crisis
of confidence. Carlos continued to reassure me, and as the wedding
details demanded more and more attention, the sense of panic
gradually faded into the background.

Our guests started arriving several days before the wedding.
Carlos's grandparents, in their mid-eighties, flew in from Argentina, as did one of his aunts and uncles. My ninety-year-old grandmother, with whom I am very close, was also able to make the
trip from California. Many more of our friends than we expected
were able to join us as well. Of our medical school friends, prac-
tically the only one missing was Roy, who was traveling in Turkey
at the time.

Our wedding felt magical to us. While we had been warned
that it would slip by before we noticed it passing—"All I have to
say is, make sure you have a good videographer," Carlos's uncle
advised—we felt that we were able to experience the entire evening. It was only afterward that the wedding began to feel unreal.
We left the following afternoon for Ireland. On the plane Carlos
kept repeating, "Here we are on our honeymoon, El. Our *honeymoon!*"

We rented a car in Ireland and spent two weeks driving around the countryside. The spring hills were covered with wild rhodo-dendrons and fields of yellow irises. After the busyness of the weeks preceding our wedding, we were able to relax and enjoy being together. But by the last few days of our honeymoon, we both started to feel the anxiety of impending internship again. I dreamed that I failed an examination that determined if I would be a good doctor, and I had nightmares about physicians berating me for my gross incompetence. Carlos dreamed that he and I were the only two people available to care for a child in the midst of a severe asthma attack.

We returned to Boston just two days before our medical school graduation. As a result, we missed nearly all of the pregraduation class events and attended only the graduation itself.

It was hard to shift gears from the wedding to graduation. Neither Carlos nor I was particularly excited about graduation. In some ways I was surprised. I thought back to first year, when I lived in Vandy, the dorm immediately across the street from the medical quadrangle where graduation was held. I could hear the ceremony from my room, and as I looked out my window, I had seen graduates and their families walking back and forth all af-ternoon. Acutely aware of the hurdles that still lay before me until I would reach that day, I had been envious of those grad-uates. My eyes had smarted a little in anticipation of the tears of joy I would weep when I graduated.

But here I was at graduation day. When our alarm went off at five forty-five, I could barely drag myself out of bed. Carlos's parents, who stayed with us, despite their California jet lag, were much quicker to get up, and by seven twenty-five when we ar-rived at the undergraduate campus for the morning ceremony, my parents were already waiting in line to get the best seats. Carlos and I left our parents and went to line up with our class-mates for the morning processional. We waited around in semi-organized chaos for two hours until we marched into the Harvard Yard, where the ceremony was held.

As we stood talking, many of us thought the moment was anti-climactic. "I've already been on vacation for four weeks. This heralds the beginning of residency in two weeks more than a release from medical school," Roy said. "If I felt ready to be a doctor, I'd be a lot more excited right now."

After marching into Harvard Yard, we sat near students from the other graduate schools, and we traded cheers as we waited for the ceremony to begin. The graduates of the school of education yelled, "We will, we will, school you! SCHOOL YOU!" Several of our classmates responded, "We will, we will, heal you! HEAL YOU!" Some people sitting behind me amended the cheer: "We will, we will, KILL YOU! KILL YOU!" We laughed. Even at the moment of our graduation our feelings of inadequacy emerged.

In the afternoon there was a private ceremony for the medical and dental school graduates back at the medical school. A tent had been set up in the green at the center of the medical quadrangle. After lunch all 157 medical and dental graduates lined up on the steps of the medical school for our class photograph. We were arranged alphabetically to be in the correct order to receive our diplomas. Nearly half our classmates had chosen to take an extra year, and nearly half the people graduating with us had taken time off and were returning from previous years. I read the class list diagram to see where I needed to line up, and I didn't recognize any of the names of the people standing near me.

We all had gone our separate ways after second year. I didn't even remember the names of some of my classmates anymore. As I sat waiting for my name to be called, surrounded by people I didn't know, I felt it was artificial to come together after four years and expect to share a meaningful group experience.

For some of my classmates, however, graduation truly was an exhilarating moment. Several of our classmates were older and had come to medicine after pursuing other careers. One in particular was in her late thirties when she entered medical school. I can only imagine what courage it must have taken to give up her career to pursue such a long, expensive, and exhausting medical education. Beaming, she carried a banner as she walked across the stage to receive her diploma: "Thank you, Mom, Dad, and the Love of My Life, Tom." Several other classmates carried their infants or walked with their older children to receive their diplomas. Each of these young "graduates" received a stuffed animal as their parents received their diplomas.

I watched from several rows back as Carlos's name was called and he received his diploma. I felt so far away from him as I waited my turn. Finally, after what felt like an eternity, my name was called. I had decided to use my married name on my diploma (I took Carlos's surname as my middle name), and it was powerful to answer to my new name for the first time.

After I stepped off the podium, I started to walk back to my seat. Somehow, at that moment, I felt particularly depressed. But then I saw Carlos standing in the aisle just in front of my row. Instead of returning to his seat as expected after receiving his diploma, he had waited to greet me after I received mine. I was so touched by his gesture, and now the tears came that I had anticipated when I overheard graduation as a first-year student.

Afterward my mother said, "Well, maybe it was anticlimactic for you, but it was very meaningful for us as parents. Somehow watching you and Carlos walking hand in hand in your matching black robes and pink hoods was very powerful for us. It marked the end of an era."

Now my degree had been conferred and I was officially a doctor. But as I reflected on the past four years, I wondered when that transformation had occurred. At what moment did I feel like a doctor?

Was it during our white coat ceremony on the first day of first year, when we received our white coats with "Harvard Medical School" emblazoned in crimson cursive over our left breast? Did it happen during first year while watching *ER*, when I first began to understand the medical language? Or perhaps it happened during second year, when I bought my medical equipment in preparation for learning the physical exam. I felt certain then that I was on the brink of learning what I had come to medical school to do: patient care. When I began third year, I thought surely the transition would happen now that I entered the wards.

Only two months before graduation, in the emergency department, I had experienced my most devastating moment in medicine, the moment that caused me to doubt my choice to become a doctor. I had confronted my fallibility. During the intervening months, with Carlos's support, I had gained a measure of perspective on the incident. I can only be human, and I had to forgive myself for that. While my self-confidence was still a little shaky, I finally scraped together the courage to move on. I regained my conviction that I could and would be a good doctor.

After four years, my white coat finally feels familiar on my shoulders. Now grayed with use, one of the pockets has torn and is secured with staples as a temporary measure. I have grown accustomed to the rhythm of life in the hospital. Once awed by the medical intricacies depicted on *ER*, I now readily pick out the inaccuracies. Do I feel like a doctor yet? I'm still not sure. I suspect I will never finish growing into my role as doctor and caregiver.